Springer Theses

Recognizing Outstanding Ph.D. Research

Aims and Scope

The series "Springer Theses" brings together a selection of the very best Ph.D. theses from around the world and across the physical sciences. Nominated and endorsed by two recognized specialists, each published volume has been selected for its scientific excellence and the high impact of its contents for the pertinent field of research. For greater accessibility to non-specialists, the published versions include an extended introduction, as well as a foreword by the student's supervisor explaining the special relevance of the work for the field. As a whole, the series will provide a valuable resource both for newcomers to the research fields described, and for other scientists seeking detailed background information on special questions. Finally, it provides an accredited documentation of the valuable contributions made by today's younger generation of scientists.

Theses may be nominated for publication in this series by heads of department at internationally leading universities or institutes and should fulfill all of the following criteria

- They must be written in good English.
- The topic should fall within the confines of Chemistry, Physics, Earth Sciences, Engineering and related interdisciplinary fields such as Materials, Nanoscience, Chemical Engineering, Complex Systems and Biophysics.
- The work reported in the thesis must represent a significant scientific advance.
- If the thesis includes previously published material, permission to reproduce this must be gained from the respective copyright holder (a maximum 30% of the thesis should be a verbatim reproduction from the author's previous publications).
- They must have been examined and passed during the 12 months prior to nomination.
- Each thesis should include a foreword by the supervisor outlining the significance of its content.
- The theses should have a clearly defined structure including an introduction accessible to new PhD students and scientists not expert in the relevant field.

Indexed by zbMATH.

Lorenzo Piro

Optimal Navigation in Active Matter

Doctoral Thesis accepted by
the Georg August University of Göttingen, Göttingen,
Germany

Springer

Author
Dr. Lorenzo Piro
Department of Living Matter Physics
Max Planck Institute for Dynamics
and Self-Organization (MPI-DS)
Göttingen, Niedersachsen, Germany

Supervisors
Prof. Ramin Golestanian
Department of Living Matter Physics
Max Planck Institute for Dynamics
and Self-Organization (MPI-DS)
Göttingen, Niedersachsen, Germany

Rudolf Peierls Centre for Theoretical
Physics
University of Oxford
Oxford, UK

Dr. Benoît Mahault
Department of Living Matter Physics
Max Planck Institute for Dynamics
and Self-Organization (MPI-DS)
Göttingen, Niedersachsen, Germany

ISSN 2190-5053 ISSN 2190-5061 (electronic)
Springer Theses
ISBN 978-3-031-52576-6 ISBN 978-3-031-52577-3 (eBook)
https://doi.org/10.1007/978-3-031-52577-3

This Springer imprint is published by the registered company Springer Nature Switzerland AG
The registered company address is: Gewerbestrasse 11, 6330 Cham, Switzerland

Supervisors' Foreword

Biological microswimmers often live on a limited energy supply, such that the optimization of biological processes offers them with a certain evolutionary advantage. In particular, efficient propulsion and navigation in terms of energy dissipation and travel time is of crucial importance for these microorganisms. The design of optimally navigating artificial swimmers also has potentially valuable applications such as targeted drug delivery, which has become an increasingly realistic perspective due to the recent progress in the experimental realization of controllable microswimmers able to perform complex tasks.

The first optimal navigation problem was formulated in 1931 by the mathematician Ernst Zermelo, who was interested in determining, for a boat sailing in windy conditions, the trajectory ensuring shortest arrival time at a given target. Since this pioneering work, much theoretical progress has been achieved while the field, however, still faces open challenges. For microswimmers navigation is notably much more challenging than at our macroscopic scale, as these microorganisms are subject to additional complexities such as fluctuations or hydrodynamic interactions.

Lorenzo's Ph.D. work precisely addresses some of these important questions, and provides a significant advance towards real world applications. The material presented in this book indeed offers valuable insights on how efficient navigation may be achieved in the presence of confining forces and flows, or thermal fluctuations. On the more conceptual side, the scientific endeavors of Lorenzo led him to draw interesting connections between the problem of navigation on Riemannian manifolds and the geodesics of Randers spaces that were initially introduced in the context of general relativity. In an attempt to establish more realistic theoretical descriptions, the energetic cost of navigation is finally addressed via a novel formulation of the problem that accounts for the hydrodynamic coupling between the surrounding flow field and the swimmer body geometry.

In addition, this book proposes a pedagogical introduction to the current state of the art in the theoretical approaches of navigation problems. I thus have no doubt that it will be much appreciated by graduate students and more experienced researchers with an interest in this field.

Göttingen, Germany
October 2023

Prof. Ramin Golestanian
Dr. Benoît Mahault

Abstract

Motile active matter systems are composed by a collection of agents, each of which extracts energy from the surrounding environment in order to convert it into self-driven motion. At the microscopic scale, however, directed motion is hindered by both the presence of stochastic fluctuations. Living microorganisms therefore had to develop simple yet effective propulsion and steering mechanisms in order to survive.

We may turn the question of how these processes work in nature around and ask how they should work in order to perform a task in the theoretically optimal way, an issue which falls under the name of the *optimal navigation problem*. The first formulation of this problem dates back to the seminal work of E. Zermelo in 1931, in which he addressed the question of how to steer a ship in the presence of an external stationary wind so as to reach the destination in the shortest time.

Despite the considerable progress made over the years in this context, however, there are still a number of open challenges. In this thesis, we therefore aim to generalize Zermelo's solution by adding more and more ingredients in the description of the optimal navigation problem for microscopic active particles.

First, borrowing theoretical tools from differential geometry, we here show how to extend the analytical solution of this problem to when motion occurs on curved surfaces and in the presence of arbitrary flows. Interestingly, we reveal that it can elegantly be solved by finding the geodesics of an asymmetric metric of general relativity, known as the *Randers metric*.

Then, we study the case in which navigation happens in the presence of strong external forces. In this context, route optimization can be crucial as active particles may encounter trapping regions that would substantially slow-down their progress. Comparing the exploration efficiency of Zermelo's solution with a more trivial strategy in which the active agent always points in the same direction, here we highlight the importance of the optimal path stability, which turns out to be fundamental in the design of the proper navigation strategy depending on the task at hand.

We then take it a step further and include a key ingredient in the comprehensive study of optimal navigation in active matter, namely stochastic fluctuations. Although methods already exist to obtain both analytically and numerically the optimal strategies even in the presence of noise, their implementation requires the presence of an external interpreter that takes away the active agent's autonomy. Inspired by the tactic behaviours observed in nature, we here introduce a whole new class of navigation strategies that allows an active particle to navigate semi-autonomously in a complex and noisy environment. Moreover, our study reveals that the performance of the theoretical optimal strategy can be reproduced starting from some simple principles based on symmetry and stability arguments.

Finally, we lay the ground for moving towards a more realistic description of the problem. In fact, we extend the optimization problem by also considering the energetic costs involved in navigation and how these depend on the shape of the active particle itself. Remarkably, our analysis uncovers the existence of an interesting trade-off between the minimization of the arrival time at a target and the corresponding energetic cost, which in turn determines the optimal shape of an active particle.

Acknowledgements

During these three years as a Ph.D. student in the department of Living Matter Physics (LMP) in Göttingen, I have been faced with a number of challenges. Nevertheless, through thick and thin, I have had the huge pleasure of getting to know many wonderful people who have contributed, each in their own way and to a different extent, to my achievement of this milestone.

First and foremost, I would like to thank my supervisor Ramin Golestanian for diligently guiding me through this journey with his constant support. I am sincerely grateful to him for always inspiring me to improve my skills as young researcher and to do my best with his valuable advice and critical comments.

This work would never have reached its current form without the careful and critical revisions of my group leader Benoît Mahault. Over all these years, his door has always been open for me, relentlessly willing to provide me with the help I needed to move forward, especially in my moments of greatest frustration. I learnt more from him about this job than I could ever have imagined, and for that (and his endless patience) I am eternally grateful.

I acknowledge our collaborators Evelyn Tang and Andrej Vilfan for their insightful contributions that led to the development of the ideas behind the results presented in Chaps. 2 and 5 of this thesis, respectively.

I also thank the members of my thesis advisory committee (TAC) Jörg Enderlein and Matthias Krüger for their helpful inputs and feedback during our meetings. Furthermore, I would like to thank Stefan Klumpp, Viola Priesemann and Aljaž Godec for showing interest in my work and accepting to be part of the thesis examination board.

One of the challenges in my Ph.D. has been navigating around the hurdles of bureaucracy. I would therefore like to thank Ayşe Bolik, Viktoryia Novak, Antje Erdmann and Frauke Bergmann for all the administrative and technical support they have tirelessly and patiently given me throughout these years.

Then, I would like to thank all my colleagues in the LMP department. On the one hand, the stimulating and (literally) active environment in the department kept my passion and interest in scientific research alive. On the other hand, I feel blessed to have been able to work with people who are, for the most part, my dear friends. I will never forget the happy hours, the parties, the volleyball matches (*Go Le Rane!*), the countless pints of beers we drank and schnitzels we ate together. I would also like to thank all my friends outside the department that I was lucky enough to meet here in Göttingen. Thanks to all of you I could always find a way to relieve the stress from work and recharge the batteries before the start of another week.

A special thanks goes to Vincent Ouazan-Reboul. Together we have been through a lot and, whatever the circumstances, I could always count on him to offer me the back-up I sought both within and beyond work. I could not have wished for a better person to share this whole journey with.

It goes without saying that I thank my parents, Roberto and Roberta, and my brother Daniele for always cheering for me and encouraging me in difficult times. Although from afar, their unconditional love and support have been essential to give me the guts to get over every obstacle.

Last but not least, I would like to thank Martina for consistently backing me up in my decisions and always being there for me, despite the distance. Her extraordinary strength has been my secret weapon during the Ph.D. and I can safely say that this work would never have achieved this form without her constant loving support.

Contents

Chapter 1
Introduction

If I have seen further, it is by standing on the shoulders of Giants.

Isaac Newton

1.1 Models of Active Matter Systems

A unifying characteristic of various living organisms is that they can extract and consume energy in order to move, apply mechanical forces or deform their shape [1]. Collections of such self-driven agents fall under the name of *active matter* systems [2]. Due to the constant energy dissipation at the individual level, these systems are continuously driven far from thermal equilibrium. This characteristic allows them to exhibit a variety of intriguing phenomenons without any equilibrium equivalent, such as the emergence of collective motion [3, 4], self-organized structures [5–8] or activity-induced patterns [9, 10]. A broad class of active matter systems is that characterised by agents capable of moving autonomously in their environment, hence referred to as *motile active matter* [11] which spans several scales, ranging from bacteria [12], molecular motors [13, 14] or cells [15, 16], to schools of fish, flocks of birds or even human crowds [17, 18].

To this day, there are countless open challenges in this field due to the high degree of complexity of these systems, be they living or man-made [19]. As a result, the study of active matter has recently brought together a growing number of scientists from various disciplines including materials science [20], robotics [21–23], chemical [24], biological [25], soft matter [2, 26] and statistical physics [1]. Over the years, researchers have thus been devoted to the development of new minimal models aimed at rationalizing, predicting and controlling the properties of motile active matter systems. These are typically phenomenological models based on global principles such as conservation laws and symmetries, which help to minimize the number of parameters controlling the dynamics of the system while also making them simple enough to be numerically efficient.

L. Piro, *Optimal Navigation in Active Matter*, Springer Theses,
https://doi.org/10.1007/978-3-031-52577-3_1

An outstanding example of minimal model inspired by animal behavior [27, 28] is the renowned *Vicsek model* [29], which was proposed to understand the collective motion of a group of agents. It consists of a set of point-like particles self-propelling in two dimensions and influencing each other's orientation through local alignment of their velocities. Despite the simplicity of this model, it remarkably allows to study collective motion across many scales, spanning from microtubules [30] to flocks of birds [31]. It represents, in fact, one of the simplest of models showing a transition to collective motion within a population of self-propelling agents and, in the study of active matter, it certainly plays the role of prototype in a way analogous to Ising's model in ferromagnetism.

Similarly, there has been a growing interest in the modeling of microscopic self-propelled organisms at the individual level, also known as *microswimmers* or *active particles* [32]. The former term refers to force-free and torque-free organisms featuring a hydrodynamic coupling with the fluid they are immersed into, which occurs via the force fields generated by their swimming patterns. The latter defines the broader class of objects moving in an inert medium (typically a viscous fluid) yielding just hydrodynamic drag and a stochastic exchange of momentum.[1] In either case, these microorganisms are able to draw energy from their surroundings and then turn it into directed motion by means of a variety of mechanisms, which we shall now briefly review.

1.1.1 Self-propulsion Mechanisms

Directed motion at the microscale is especially problematic because of the combination of two factors: the absence of inertia and the presence of thermal fluctuations. On those scales, microorganisms indeed live in a world where fluid friction and viscosity dominate over inertia, also known as the low Reynolds number (or *overdamped*) regime [33–36]. The absence of inertia has interesting and counter-intuitive implications, as illustrated by the *scallop theorem* [36]: a swimmer making a time-symmetric movement cannot achieve a net displacement at low Reynolds numbers. This has therefore led biological microorganisms to evolve propulsion mechanisms that break temporal reversibility.

In general, since microswimmers motion is momentum-conserving,[2] the far-field flow they generate can be well mathematically described by a force dipole [34, 35]. As shown schematically in Fig. 1.1a, this characterization splits microswimmers into two classes. *Pushers*, who move the fluid towards them on the sides of their body and

[1] Despite this difference, in the following we will actually not make any distinction between these two terms since their observable behavior—when considered individually—is indistinguishable in a homogeneous environment [32].

[2] Due to the absence of inertia at low Reynolds number.

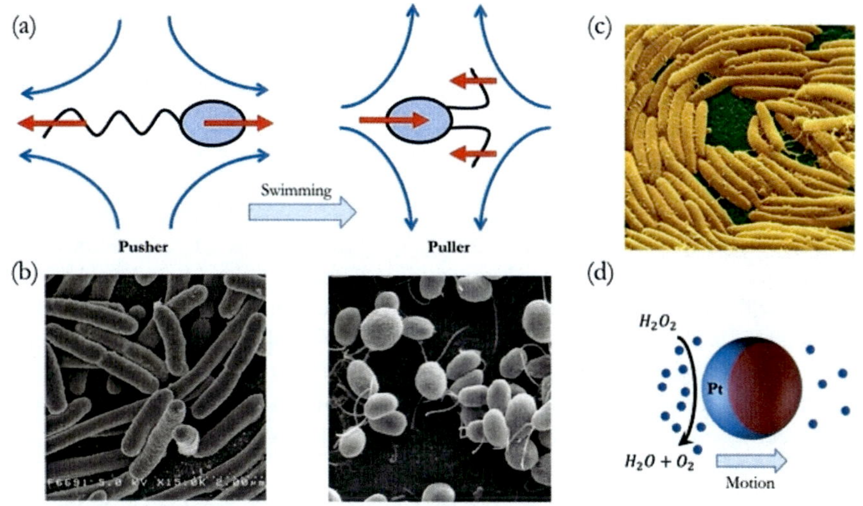

Fig. 1.1 **a** Schematic representation of two self-propulsion mechanisms of microswimmers at low Reynolds number. On the left, pushers generate the propulsion from the rear end due to the rotation of a bundle of flagella. Pullers (on the right) generate instead the force from the front end thanks to two flagella beating in a breaststroke way. Thanks to these non-reciprocal beating patterns they are able to swim in one direction (from left to right in this figure). **b** Left panel: a colony of *E. Coli* bacteria observed at the microscope as an example of pusher microswimmer. *Source* Wikimedia Commons, Rocky Mountain Laboratories, NIAID, NIH, public domain. Right panel: *Chlamydomonas reinhardtii* algae under a Scanning Electron Microscope (SEM). A paradigmatic example of pullers. *Source* Wikimedia Commons, Dartmouth Electron Microscope Facility, Dartmouth College, public domain. **c** A swarm of *Myxococcus xanthus* under a SEM, an example of gliding bacteria. *Source* ETH Zurich/Gregory J. Velicer and Jürgen Berger. **d** Illustration of the self-diffusiophoresis mechanism characterizing the motion of a Janus particle. This colloid is driven by the local concentration gradient of the reaction product (oxygen, O_2) generated by the catalysis of hydrogen peroxide (H_2O_2) on the platinum-coated side (blue hemisphere)

away from them at the extremities, and whose forces forming the dipole point away from them. An emblematic example is the *Escherichia Coli* bacterium, which uses rotating flagella at the back of its body to self-propel [37] (left panel in Fig. 1.1b). In contrast, when the forces forming the dipole point towards the microswimmer, it is called *puller*: it draws fluid towards itself in the direction of motion and pushes it out from the sides. An example is given by the *Chlamydomonas reinhardtii*, an alga with two flagella that move collecting fluid from the front as in a breaststroke style [38] (right panel in Fig. 1.1b).

However, there are many situations in which the presence of the fluid can be neglected or taken into account effectively through friction forces. This is generally the case when local interactions dominate the dynamics like, for e.g., in two-dimensional systems where the swimmers are in contact with a substrate that acts as a momentum sink and screens hydrodynamic effects. Biological examples are gliding bacteria like the *Myxococcus xanthus* (shown in Fig. 1.1c) whose motility relies on specific pili [39], and also crawling eukaryotic cells which self-propel using their actin cytoscheleton [40].

At the same time, a variety of artificial microswimmers have been experimentally realized in order to reproduce the qualitative behavior of motile biological active matter [32, 33]. One of the simplest examples is provided by self-propelling liquid droplets [41]. Their motion is caused by a Marangoni flow [42] generated by a self-sustained gradient in the surface tension of the droplet. Experimentally, this effect has been shown for water droplets containing bromine in an oil suspension [43]. Another paradigmatic example is provided by Janus particles [44–46], spherical colloids with hemispheres of distinct physicochemical properties. These particles are typically half-coated with a layer of platinum (Pt) that catalyzes a chemical reaction in an aqueous solution. As schematically shown in Fig. 1.1d, this in turn breaks the symmetry of the reaction product distribution around the particle, giving rise to phoretic and osmotic forces [47, 48]: a propulsion mechanism known as *self-diffusiophoresis*.

All in all, a microswimmer can autonomously accomplish directed motion in several ways. We shall now discuss some minimal models that effectively describe the generic features of the motion of such microscopic active particles regardless of their specific propulsion mechanism.

1.1.2 Models of Active Particles

As anticipated at the beginning of the previous section, a particle immersed in a fluid is subject to the stochastic forces resulting from collisions with the fluid molecules. The corresponding motion that these interactions generate is called *Brownian motion*, named after the botanist Robert Brown who first noticed and described this phenomenon in 1827 [49]. The intensity of these fluctuations can be quantified via the so-called translational diffusion coefficient D which depends both on the fluid temperature T and on the particle mobility μ via the renowned Einstein relation [50]: $D = \mu k_B T$, with k_B being the Boltzmann constant. At the same time, the particle also undergoes rotational diffusion over time scales τ_r given by the inverse of the corresponding rotational diffusion coefficient D_r.

Thus, we can now write down the overdamped equations of motion of a passive particle immersed in a quiescent fluid in two dimensions as[3]:

$$\begin{cases} \dot{\boldsymbol{r}} = \sqrt{2D}\boldsymbol{\xi} \\ \dot{\theta} = \sqrt{2D_r}\xi_\theta \ , \end{cases} \tag{1.1}$$

[3] The generalization to the three-dimensional case is straightforward, with the particle position represented by Cartesian coordinates (x, y, z) and the orientation by the azimuthal and polar angles (ϕ, θ).

where $\boldsymbol{r} = x\,\hat{\boldsymbol{x}} + y\,\hat{\boldsymbol{y}}$ is the particle position, θ its orientation angle, while $\boldsymbol{\xi}$ and ξ_θ stand for independent delta-correlated white noises with unit variance, used to model the randomness of thermal fluctuations. These stochastic dynamics represent a model for what is known as a *passive Brownian particle* (PBP) whose motion is purely diffusive and each degree of freedom is completely decoupled from the others. The overdamped dynamics in Eq. (1.1) moreover implies that the average particle displacement with respect to its initial position will always vanish. These random fluctuations therefore hinder the ability of a microswimmer to perform an effective directed motion even in a quiescent fluid. It is then quite crucial for an active particle to find a good steering strategy that can overcome thermal fluctuations.

In this regard, there exists models with various degrees of sophistication to describe active motion [51, 52]. However, here we will ignore the specific self-propulsion mechanisms and rather focus on a generic description. To this end, self-propelling particles can be divided into two main classes. On the one hand, the so-called *active Brownian particles* (ABPs). In this case the microswimmer self-propels at a constant speed v_0, but its orientation gradually changes due to rotational diffusion, which causes a non-trivial coupling between rotation and translation. The corresponding overdamped equations of motion in 2D are

$$\begin{cases} \dot{\boldsymbol{r}} = v_0\hat{\boldsymbol{u}} + \sqrt{2D}\boldsymbol{\xi} \\ \dot{\theta} = \sqrt{2D_r}\xi_\theta \ , \end{cases} \tag{1.2}$$

where $\hat{\boldsymbol{u}} = \cos\theta\,\hat{\boldsymbol{x}} + \sin\theta\,\hat{\boldsymbol{y}}$ is the unit vector corresponding to the particle intrinsic direction of motion. Note that for active particles, in general, noise could result from sources other than thermal. Activity itself can in fact give rise to stochastic fluctuations. A clear example are the chemical reactions underlying the self-propulsion mechanisms of Janus particles, whose motion observed in experiments is remarkably well reproduced by this simple effective model [44, 53]. In Fig. 1.2a are shown some sample trajectories of ABPs with different self-propelling speeds v_0 obtained from numerical simulations of (1.2).

On the other hand, we find the so-called *run-and-tumble particles* (RTPs). As the name suggests, these microswimmers switch between two states: runs, where they move at constant speed v_0 in a straight line, and tumbles, when they abruptly and randomly change their orientation. As a result, RTPs differ from ABPs in their orientation dynamics [54]. Namely, the tumbling events of a RTP are uncorrelated, occur at an average rate α and their number in a certain time window has a Poisson distribution, such that the tumbling probability is given by $P_{\text{tumble}} = 1 - e^{-\alpha}$. The orientation dynamics thus becomes [55]

$$\dot{\theta} = \sum_i \Delta\theta_i \delta(t - \tau_i) \ , \tag{1.3}$$

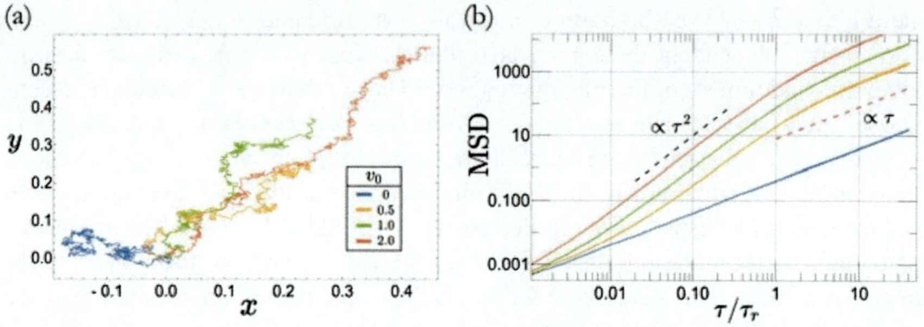

Fig. 1.2 **a** Four exemplary trajectories of ABPs with different self-propulsion speeds v_0 starting from $\boldsymbol{r}_0 = \boldsymbol{0}$ and with an initial orientation $\theta_0 = \pi/4$. The translational and rotational diffusion coefficients are here set to $D = 0.01$ and $D_r = 0.1$, respectively. **b** Corresponding curves of the mean square displacement (MSD) obtained from numerical simulations taking an ensemble average over 10^3 trajectories. You can notice the three different scaling regimes that an ABP undergoes: diffusive at very short times, ballistic when $\tau/\tau_r \approx 1$ and again diffusive at times $\tau/\tau_r \gg 1$. Legend and parameters as in **a**

where the sum runs over all the tumbling events occurred at times τ_i, while the angle variation $\Delta\theta_i$ is drawn from a uniform distribution between 0 and 2π. This model of active particles has been developed to describe the motion of *E. Coli* bacteria [12] and *Chlamydomonas algae* [56] whose propulsion mechanisms have been already discussed above.

Although the dynamics at short times depend on the model adopted, both ABPs and RTPs share the same properties at long times. It is possible to quantify this by looking at the mean square displacement (MSD) of an active particle, which measures how much on average it departs from its initial position $\boldsymbol{r}_0$ within a time τ, i.e. mathematically defined as $\mathrm{MSD}(\tau) = \langle(\boldsymbol{r}(\tau) - \boldsymbol{r}_0)^2\rangle$, where $\langle ... \rangle$ stands for the ensemble average. For an active particle, be it an ABP or a RTP, it is given by [57]

$$\mathrm{MSD}(\tau) = (4D + v_0^2\tau_r)\tau + \frac{v_0^2\tau_r^2}{2}(e^{-\frac{2\tau}{\tau_r}} - 1)\,, \tag{1.4}$$

where τ_r is the typical time scale of rotational diffusion, such that $\tau_r = D_r^{-1}$ or $\tau_r = \alpha^{-1}$ for ABPs or RTPs, respectively. Depending on the observed time scale, we can distinguish three different regimes, as shown in Fig. 1.2b. In the limit $\tau \ll \tau_r$, (1.4) reduces to $\mathrm{MSD}(\tau) = 4D\tau$, which is a purely diffusive motion with diffusion constant D and is also the exact solution for the MSD of a PBP. However, at slightly longer times, i.e. when $\tau \approx \tau_r$, the motion is effectively ballistic since $\mathrm{MSD}(\tau) = 4D\tau + 2v_0^2\tau^2 \propto \tau^2$. Lastly, at time scales much larger than the characteristic rotational diffusion, i.e. $\tau \gg \tau_r$, we get $\mathrm{MSD}(\tau) = (4D + 2v_0L)\tau \propto \tau$, such that we recover again a diffusive motion with an enhanced diffusion coefficient $D_{\mathrm{eff}} = D + v_0L/2$. We have here defined $L \equiv v_0\tau_r$, known as the active particle persistence length, i.e. the length scale over which it travels straight on average before its direction is randomized.

There are, however, *limiting* regimes in which the impact of noise is negligible. A good measure is provided by the Péclet number (Pe) of the system. This dimensionless parameter quantifies the relative importance between directed motion and diffusive effects, and for an ABP can be defined as [32]

$$\mathrm{Pe} \propto \frac{v_0}{\sqrt{D_r D}}, \tag{1.5}$$

up to a constant numerical prefactor. In cases where Pe is large, the effects of stochastic fluctuations on the motion of the microswimmer can be neglected. These essentially correspond to situations in which the characteristic time scale of diffusion is much larger than the observed time window. In practice, this is the case when, e.g., active motion occurs in a high-viscosity fluid, such as honey, oils or long-chain hydrocarbons [58, 59].

We have thus shown that the motion of an active particle is characterised by both a ballistic regime at relatively short times and a diffusive one at longer times. Moreover, the transition between the two regimes turns out to be essentially governed by the swimmer persistence length, which in turn depends on its intrinsic activity. In practical terms, the ability of tuning this quantity can thus be crucial for the survival of biological microswimmers navigating in a complex environment [12, 60, 61].

1.1.3 Smart Active Particles

Through millions of years of evolution, biological microswimmers had to develop mechanisms to control their own activity in order to accomplish vital tasks. Some examples are the search strategies displayed by bacteria like *E. Coli* while looking for their nutrient [12, 34] or by spermatozoa trying to locate the egg [62]. Through the comparison of the chemical concentration levels along their path, these microorganisms are capable of altering their tumbling rate over time so as to effectively climb up the concentration gradient. This process is called *klinokinesis with adaptation* [63] and results in a biased random walk known as *chemotaxis* [64], which is depicted in Fig. 1.3.[4]

However, these adaptation mechanisms—first observed over 100 years ago [66]—characterize only a specific class of microswimmers since they require sophisticated sensory and motility machinery as well as internal information processing, which in turn involves memory. On the other hand, behaviors observed in nature have been a source of inspiration for the design of artificial self-propelled microswimmers capable of performing autonomously specialized tasks in complex environments [67, 68] in a much simpler fashion.

---

[4] An analogous behavior has also been observed in photosynthetic biological microorganisms in the presence of light gradients and is thus called *phototaxis* [65].

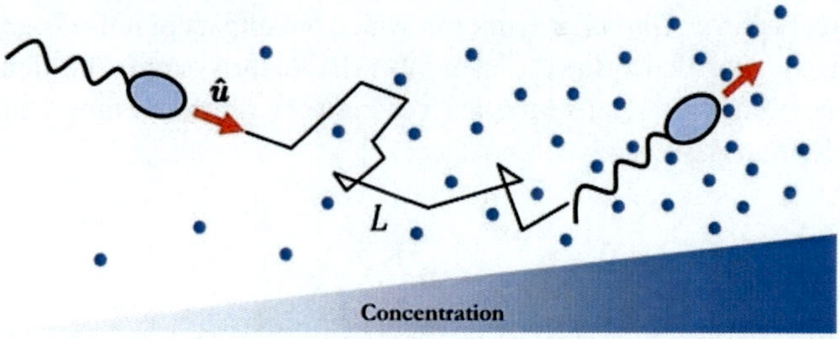

Fig. 1.3 Graphical representation of bacterial chemotaxis. By self-regulating its own motility apparatus, and thereby its persistence length L, a microswimmer is able to climb up chemical concentration gradients autonomously. Here, the red arrow represents the swimmers heading direction $\hat{\boldsymbol{u}}$ while the blue circles stand for the chemical particles, whose concentration grows from right to left as indicated by the bar gradient underneath. Qualitatively, something similar occurs in all other types of taxis, the main difference being in the sensory and motility machinery of the active particle at hand

An example is provided by chemically active particles like, e.g., synthetic phoretic colloids, which can exhibit (anti-)chemotaxis by (anti-)aligning their axis of orientation to the local gradient of (possibly self-produced) chemicals [69–72]. Remarkably, this reproduces well the behavior displayed by individual enzymes [8, 73].

Based on this simple mechanism—known as *orthokinesis*—it has therefore recently become possible to design synthetic microswimmers capable of responding to a variety of external stimuli such as light [74–76] or viscosity [77] gradients or even gravitational [78] or magnetic [79–81] fields.

Currently, a major challenge in active matter is understanding how such processes can be optimized [82, 83], both at the collective scale and at the individual active particle level.

1.2 Optimal Control Theory

The proper mathematical framework for rigorously studying optimization problems is known as *optimal control theory* [84]. Building on the pioneering works by Richard Bellman [85] and Lev Pontryagin [86] from last century, this theoretical tool has become very popular also outside the mathematical sciences thanks to its many applications in engineering [87, 88] and finance [89]. Optimal control (OC) is essentially an extension of variational calculus aimed at finding the control strategy that minimizes/maximizes a given objective functional, also known as *cost function*.

Let us now briefly illustrate the main ideas that lead to the formal solution [90]. Consider the deterministic dynamics of an active agent generally described by the following n-dimensional differential equation

$$\dot{\boldsymbol{q}} = \boldsymbol{F}[\boldsymbol{q}(t), \boldsymbol{c}(t), t] , \tag{1.6}$$

where t indicates the time and $\boldsymbol{q}$ represents the state of the agent. The variable $\boldsymbol{c}$ is the so-called *control* upon which one can act to manipulate the state $\boldsymbol{q}$ and, in fact, it does not have a predefined equation of motion. The system (1.6) is furthermore equipped with the boundary conditions

$$\begin{cases} \boldsymbol{q}(t_0) = \boldsymbol{q}_0 \\ q_i(t_\mathrm{f}) = q_{i,f} \quad \text{for} \quad i = 1, \ldots, k \,, \end{cases}$$

where $k \leq n$, while t_0 and t_f are the initial and the final times respectively. In the following, we will treat the more general case in which t_f is not specified.

The optimization problem can actually be addressed with tools similar to those used in analytical mechanics. Let us therefore consider a cost function of the form

$$C = \phi(\boldsymbol{q}(t_\mathrm{f}), t_\mathrm{f}) + \int_{t_0}^{t_\mathrm{f}} \mathcal{L}[\boldsymbol{q}(t), \boldsymbol{c}(t), t]\,\mathrm{d}t \,, \tag{1.7}$$

where ϕ and $\mathcal{L}$ are known as the *endpoint* cost and the *running* cost respectively, although the latter can be identified as a Lagrangian. The goal is now to determine the history of $\boldsymbol{c}(t)$ that would minimize/maximize C. To this end, we should first adjoin the equations of motion (1.6) to the Lagrangian introducing a vector of Lagrange multipliers $\boldsymbol{p}$, whose components are the generalized momenta conjugated to the state variable $\boldsymbol{q}$:

$$\tilde{C} = \phi(\boldsymbol{q}(t_\mathrm{f}), t_\mathrm{f}) + \int_{t_0}^{t_\mathrm{f}} \{\mathcal{L}[\boldsymbol{q}(t), \boldsymbol{c}(t), t] + \boldsymbol{p}(t) \cdot (\boldsymbol{F}[\boldsymbol{q}(t), \boldsymbol{c}(t), t] - \dot{\boldsymbol{q}})\}\,\mathrm{d}t \,. \tag{1.8}$$

We may identify the system Hamiltonian as

$$\mathcal{H}[\boldsymbol{q}(t), \boldsymbol{c}(t), t] \equiv \mathcal{L}[\boldsymbol{q}(t), \boldsymbol{c}(t), t] + \boldsymbol{p}(t) \cdot \boldsymbol{F}[\boldsymbol{q}(t), \boldsymbol{c}(t), t] \,, \tag{1.9}$$

and then integrate by parts the last term in (1.8), yielding

$$\tilde{C} = \phi + [\boldsymbol{p} \cdot \boldsymbol{q}]_{t_0} - [\boldsymbol{p} \cdot \boldsymbol{q}]_{t_\mathrm{f}} + \int_{t_0}^{t_\mathrm{f}} (\mathcal{H} + \dot{\boldsymbol{p}} \cdot \boldsymbol{q})\,\mathrm{d}t \,,$$

where we have dropped the explicit dependencies. Let us now compute the variation in the cost $\tilde{C}$, which reads

$$\delta\tilde{C} = [(\partial_t\phi + \mathcal{H})\mathrm{d}t_\mathrm{f} + (\nabla_{\boldsymbol{q}}\phi - \boldsymbol{p}) \cdot \mathrm{d}\boldsymbol{q}]_{t_\mathrm{f}} + \int_{t_0}^{t_\mathrm{f}} [(\nabla_{\boldsymbol{q}}\mathcal{H} + \dot{\boldsymbol{p}})\delta\boldsymbol{q} + \nabla_{\boldsymbol{c}}\mathcal{H}\delta\boldsymbol{c}]\,\mathrm{d}t \,,$$

where we have used the identity $\delta\boldsymbol{q}(t_\mathrm{f}) = \mathrm{d}\boldsymbol{q}(t_\mathrm{f}) - \dot{\boldsymbol{q}}(t_\mathrm{f})\mathrm{d}t_\mathrm{f}$ and placed $\delta\boldsymbol{q}_0 = \boldsymbol{0}$, since $\boldsymbol{q}(t_0)$ is fixed. At optimality, the cost function has to be stationary, i.e. $\delta\tilde{C} = 0$. By requiring this to hold for arbitrary variations of $\boldsymbol{q}$, $\boldsymbol{c}$ and t_f, we obtain the necessary conditions for optimality, also known as Pontryagin's principle [86]:

$$
\begin{cases} \dot{\boldsymbol{p}} = -\nabla_{\boldsymbol{q}}\mathcal{H} \\ \nabla_{\boldsymbol{c}}\mathcal{H} = 0 \\ [\partial_t \phi + \mathcal{H}]_{t_\mathrm{f}} = 0\,, \end{cases} \tag{1.10}
$$

together with the boundary conditions:

$$
\begin{cases} \boldsymbol{q}(t_0) = \boldsymbol{q}_0 \\ q_i(t_\mathrm{f}) = q_{i,f} \text{ (specified)}; \quad i = 1, \ldots, k \\ p_i(t_\mathrm{f}) = \partial_{q_i}\phi\,; i = k+1, \ldots, n\,, \end{cases} \tag{1.11}
$$

where n is the dimensionality of the system (1.6). These last two systems, along with the equation of motion (1.6), make up the two-point boundary-value problem to be solved in order to obtain the OC strategy.[5]

1.2.1 Stochastic Optimal Control

The formalism presented in the previous section does not take into account the possible presence of stochastic fluctuations affecting the system's dynamics. These in fact characterize motion at the microscopic scale and are therefore a key ingredient to be included in a thorough treatment of optimization problems in active matter. To this end, however, one must resort to a slightly different approach, which we illustrate in the following.

Let us consider a system governed by the following stochastic differential equations

$$
\dot{\boldsymbol{q}} = \boldsymbol{F}[\boldsymbol{q}(t), \boldsymbol{c}(t), t] + \sqrt{2D}\boldsymbol{\xi}\,, \tag{1.12}
$$

with D being a diffusion constant and $\boldsymbol{\xi}$ a n–dimensional white noise delta-correlated process with unit variance. We will hereafter assume that the initial state of the system is fixed $\boldsymbol{q}(t_0) = \boldsymbol{q}_0$ and the final time is unconstrained. The aim here is to find the control $\boldsymbol{c}(t)$ that minimizes

$$
C(\boldsymbol{q}_0, \boldsymbol{c}) = \left\langle \phi(\boldsymbol{q}(t_\mathrm{f})) + \int_{t_0}^{t_\mathrm{f}} \mathcal{L}[\boldsymbol{q}(t), \boldsymbol{c}(t), t]\,\mathrm{d}t \;\middle|\; \boldsymbol{q}(t_0) = \boldsymbol{q}_0 \right\rangle, \tag{1.13}
$$

which is analogous to the cost function (1.7) with the only difference that here we have to take an ensemble average over all trajectories starting from $\boldsymbol{q}_0$. We can now formally define the optimal cost function, also called *value function*, simply as

---

[5] Note that the last condition in system (1.10) is necessary if and only if the final time t_f is not specified.

$$\mathcal{C}(\boldsymbol{q},t) = \min_{\boldsymbol{c}(t\to t_\mathrm{f})} C[\boldsymbol{q}(t), \boldsymbol{c}(t \to t_\mathrm{f})] \tag{1.14}$$

where $\boldsymbol{c}(t \to t_\mathrm{f})$ is the control sequence between an intermediate time t and the final time t_f. Then, according to Bellman's principle of optimality [91], this function can be expressed recursively as

$$\mathcal{C}(\boldsymbol{q},t) = \min_{\boldsymbol{c}(t\to t+\mathrm{d}t)} \left\langle \mathcal{C}(\boldsymbol{q}(t+\mathrm{d}t), t+\mathrm{d}t) + \int_t^{t+\mathrm{d}t} \mathcal{L}[\boldsymbol{q}(s), \boldsymbol{c}(s), s]\,\mathrm{d}s \;\middle|\; \boldsymbol{q}(t) = \boldsymbol{q} \right\rangle . \tag{1.15}$$

The Taylor expansion of the first term on the right reads

$$\langle \mathcal{C}(\boldsymbol{q}(t+\mathrm{d}t), t+\mathrm{d}t) \,|\, \boldsymbol{q}(t) = \boldsymbol{q} \rangle \approx \mathcal{C}(\boldsymbol{q}(t), t) + \partial_t \mathcal{C}\mathrm{d}t + \boldsymbol{F} \cdot \nabla_{\boldsymbol{q}} \mathcal{C}\mathrm{d}t + D\nabla_q^2 \mathcal{C}\mathrm{d}t \;,$$

up to first order in $\mathrm{d}t$, where we have used Itô's lemma [92] $\langle \mathrm{d}\boldsymbol{q}^2 \rangle = \mathcal{O}(\mathrm{d}t)$. Plugging then this back into (1.15), dividing by $\mathrm{d}t$ both sides and taking the limit $\mathrm{d}t \to 0$ gives

$$\partial_t \mathcal{C} + \min_{\boldsymbol{c}} \left[\mathcal{L} + \boldsymbol{F} \cdot \nabla_{\boldsymbol{q}} \mathcal{C} + D\nabla_q^2 \mathcal{C} \right] = 0 \;, \tag{1.16}$$

which is the stochastic version of the so-called *Hamilton-Jacobi-Bellman* (HJB) equation [93] with boundary condition $\mathcal{C}(\boldsymbol{q}, t_\mathrm{f}) = \phi(\boldsymbol{q}(t_\mathrm{f}))$. Finally, the minimization with respect to the control $\boldsymbol{c}$ of the expression in square brackets is carried out by imposing the optimality condition [94]

$$\nabla_{\boldsymbol{c}} \left[\mathcal{L} + \boldsymbol{F} \cdot \nabla_{\boldsymbol{q}} \mathcal{C} + D\nabla_q^2 \mathcal{C} \right] = 0 \,. \tag{1.17}$$

Thus, inserting the expression of the control $\boldsymbol{c}$ obtained from (1.17) into (1.16), we get the n-dimensional partial differential equation to be solved in order to address the stochastic optimal control (SOC) problem.

Note that this approach can also be used to obtain the OC strategy in the deterministic case by simply setting $D = 0$. This would therefore be an equivalent alternative to employing Pontryagin's principle (1.10) introduced above. There, one needs instead to solve a set of $2n$ ordinary differential equations with two-point boundary conditions. Depending on the situation, one shall decide which approach is more convenient on a case-by-case basis.

This equivalence can be readily worked out by identifying the value function gradients, $\nabla_{\boldsymbol{q}} \mathcal{C}$, with the generalized momenta $\boldsymbol{p}$ at $D = 0$. The system Hamiltonian (1.9) in that case indeed takes the form $\mathcal{H} = \mathcal{L} + \boldsymbol{F} \cdot \nabla_{\boldsymbol{q}} \mathcal{C}$, such that the first and second necessary conditions of Pontryagin's principle (1.10) match exactly with (1.16) and (1.17) respectively.[6] The third condition in (1.10) can instead be straightforwardly recovered from Eq. (1.16), namely $\partial_t \mathcal{C} = -\mathcal{H}$, and then considering the boundary condition $\mathcal{C}(\boldsymbol{q}, t_\mathrm{f}) = \phi(\boldsymbol{q}(t_\mathrm{f}))$.

[6] Specifically, the first correspondence can be proven by taking the gradients of both sides in (1.16).

We will now show the application of both these equivalent approaches in some practical examples.

1.3 The Problem of Optimal Navigation

Finding the fastest path towards a desired destination can be highly beneficial for living organisms tracking a food source [15, 95, 96], a potential mate [97], or escaping from toxic areas [98] or predators [99]. Moreover, most often their motion happens in the presence of a fluid flow or an external force, which can hinder their navigation. The optimal path is therefore distinct from the shortest one, making this a complex problem in the field of active matter.

At the same time, thanks to the recent theoretical and experimental advances in the design of artificial microswimmers [44, 45, 48, 51, 67, 100], addressing this issue has important technological applications ranging from environmental sustainability and monitoring [101, 102], route planning [103, 104] to targeted cargo delivery at the microscale [105–107].

Hence, it is of no surprise that this problem is one of the oldest known applications of OC theory. Its classical solution can be traced back to Ernst Zermelo's work in 1931 [108]. There, he addressed the problem of how to steer a vessel navigating at constant speed v_0 in the presence of a stationary wind $\boldsymbol{f}(\boldsymbol{r})$ so as to minimize the travel time to reach a given destination. Please refer to Fig. 1.4 for an illustration of the scenario. In this context, the equations governing the motion of the ship are

$$\dot{\boldsymbol{r}} = v_0 \hat{\boldsymbol{u}} + \boldsymbol{f}(\boldsymbol{r}) \,, \tag{1.18}$$

where $\boldsymbol{r} = x\,\hat{\boldsymbol{x}} + y\,\hat{\boldsymbol{y}}$ is the vessel position and $\hat{\boldsymbol{u}} = \cos\theta\,\hat{\boldsymbol{x}} + \sin\theta\,\hat{\boldsymbol{y}}$ its heading direction.

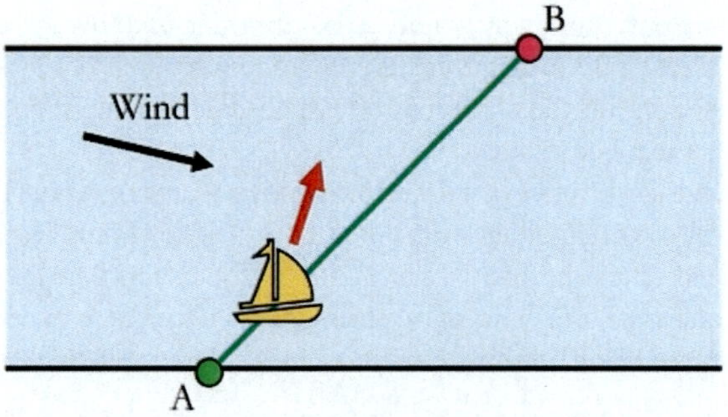

Fig. 1.4 Diagrammatic illustration of Zermelo's problem for a ship navigating under constant wind (black arrow). In order to reach the destination in the shortest time, the ship must be oriented (red arrow) so as to compensate for the transversal component of the wind. In this case indeed ZP indicates that $\theta = \theta_0$ (const), that is, the optimal path is simply straight (green line)

The solution to this problem can be readily found applying Pontryagin's principle (1.10) from OC theory [90]. Since the goal here is to minimize the total travel time, we may simply place $\mathcal{L} = 1$ and $\phi = 0$ in the cost function (1.7), such that the system Hamiltonian (1.9) takes the form

$$\mathcal{H} = 1 + \boldsymbol{p} \cdot [v_0 \hat{\boldsymbol{u}} + \boldsymbol{f}(\boldsymbol{r})] , \tag{1.19}$$

where the control variable is the ship's heading angle θ and the state is given by its position vector $\boldsymbol{r}$. In this specific case, the necessary conditions for optimality (1.10) thus translate into the following system

$$\begin{cases} \dot{p}_x = -\partial_x \mathcal{H} = -p_x \partial_x f_x - p_y \partial_x f_y \\ \dot{p}_y = -\partial_y \mathcal{H} = -\, p_x \partial_y f_x - p_y \partial_y f_y \\ 0 = \partial_\theta \mathcal{H} = v_0(-p_x \sin\theta + p_y \cos\theta) \implies \tan\theta = \frac{p_y}{p_x} \\ \mathcal{H} = 1 + \boldsymbol{p} \cdot [v_0 \hat{\boldsymbol{u}} + \boldsymbol{f}(\boldsymbol{r})] = 0 , \end{cases} \tag{1.20}$$

where the last condition holds at any time since the Hamiltonian does not depend explicitly on time and is thus a conserved quantity. We may then use the third equation to replace $\boldsymbol{p}$ by θ in this system so to obtain

$$\dot{\theta} = \sin^2\theta\, \partial_x f_y - \cos^2\theta\, \partial_y f_x + \sin\theta \cos\theta (\partial_x f_x - \partial_y f_y) , \tag{1.21}$$

which is the sought optimal steering strategy that, solved together with the equation of motion (1.18), minimizes the travel time. Hereafter, we will refer to it as *Zermelo's Policy* (ZP). One then just has to select[7] the proper initial orientation θ_0 so as to get the minimum time path that reaches the desired destination, which henceforth will be referred to as *Zermelo's path*.

For the sake of illustration, let us now consider the case where a self-propelled agent moves in the presence of a shear flow $\boldsymbol{f}(\boldsymbol{r}) = \kappa y\, \hat{\boldsymbol{x}}$ with $\kappa > 0$. Starting from the origin, the task is to reach in the shortest time another stationary point $\boldsymbol{r}_\text{T}$ at distance ℓ along the $x-$axis, i.e. $\boldsymbol{r}_\text{T} = \ell\, \hat{\boldsymbol{x}}$ with $\ell > 0$. According to Zermelo's solution, this simply amounts to solving the following set of ordinary differential equations (ODEs)

$$\begin{cases} \dot{x} = \zeta \cos\theta + y \\ \dot{y} = \zeta \sin\theta \\ \dot{\theta} = -\cos^2\theta\, . \end{cases} \tag{1.22}$$

where we have rescaled space and time as $\boldsymbol{r} \to \ell \boldsymbol{r}$ and $t \to t/\kappa$, such that the dynamics is characterized by only one non-dimensional parameter, namely $\zeta \equiv v_0/\kappa\ell$. This ODE system can be solved exactly and the corresponding solution reads

[7] This can be achieved using a shooting method [109].

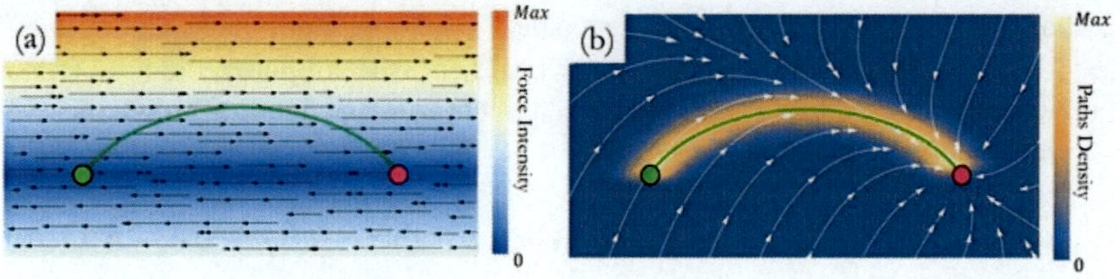

Fig. 1.5 Optimal navigation in a shear flow. **a** The ZP solution (green curve) has been obtained by integrating (1.22) at $\zeta = 1$ for an agent starting from the origin $\boldsymbol{r}_0 = \boldsymbol{0}$ (green circle) and aiming at another point downstream $\boldsymbol{r}_\text{T} = \ell\hat{\boldsymbol{x}}$ (magenta circle). The colour here codes for the flow intensity and the black arrows indicate its direction. **b** Heat map of 10^3 stochastic trajectories obtained from numerical simulations of OP in the same setup as in **a** with the diffusivity set to $D = 0.01$. The white arrows indicate the corresponding optimal control map $\hat{\boldsymbol{u}}_\text{opt}(\theta(\boldsymbol{r}))$

$$\begin{cases} x(t) = x_0 + y_0 t + \frac{\zeta}{2}\left[\text{arcsinh}[(t - \tan\theta_0)\lambda + \tan\theta_0\psi(t)] + \lambda t + (\tan\theta_0 - t)(\psi(t) - \lambda)\right] \\ y(t) = y_0 - \zeta\left[\psi(t) - \lambda\right] \\ \theta(t) = \arctan(\tan\theta_0 - t)\,, \end{cases} \tag{1.23}$$

where we have defined $\lambda \equiv \sqrt{1 + \tan^2\theta_0}$ and $\psi(t) \equiv \sqrt{1 + (t - \tan\theta_0)^2}$, with (x_0, y_0) and θ_0 being the initial position and orientation of the active agent, respectively.

In order to solve the full problem, all that is left to determine are the initial angle θ_0 and the (optimal) arrival time at the target t_opt. These depend on the relative position of the target $\boldsymbol{r}_\text{T}$ and the initial point $\boldsymbol{r}_0$. Since in our case these share the same y-coordinate, i.e. $y_T = y(t_\text{opt}) = y_0$, from (1.23) we obtain

$$t_\text{opt} = 2\tan\theta_0\,. \tag{1.24}$$

Moreover, the target position is such that $x_T - x_0 = x(t_\text{opt}) - x_0 = 1$ (in units of ℓ). This condition, together with (1.23) and (1.24), implies

$$\text{arcsinh}(2\tan\theta_0\lambda) + 2\tan\theta_0\lambda = 2/\zeta\,, \tag{1.25}$$

which is an implicit equation to be solved for θ_0. Hence, the full system of Eqs. (1.23–1.25) finally represents the complete analytical solution to the optimal navigation problem in a shear flow. The resulting Zermelo's path corresponds to the green curve shown in Fig. 1.5a. Here, the agent exploits the flow to reach the target faster. Nevertheless, it also has to be cautious not to go too far in the vertical direction to avoid being carried away by the flow, making this a non-trivial optimization problem.

Zermelo's approach, however, does not account for thermal fluctuations, which play a prominent role at the micro-scale. In such a case, one must therefore reframe the problem of optimal navigation within the context of SOC theory. To this end, let us consider a self-propelled particle moving on the plane at fixed speed v_0 in presence of a stationary force field $\boldsymbol{f}(\boldsymbol{r})$ and translational diffusion with diffusivity D. Its motion obeys the following stochastic differential equation

$$\dot{\boldsymbol{r}} = v_0\hat{\boldsymbol{u}} + \boldsymbol{f}(\boldsymbol{r}) + \sqrt{2D}\boldsymbol{\xi}\,, \tag{1.26}$$

which just corresponds to a particular choice for the dynamics in (1.12).

Since our goal is to minimize the total travel time to reach a target at position $\boldsymbol{r}_\mathrm{T}$, we may identify the particle position with the state variable and simply set $\phi = 0$ and $\mathcal{L} = 1$ in the definition of the cost function (1.13), which will thus take the form $C(\boldsymbol{r}) = \langle t_\mathrm{f} - t_0\rangle_{\boldsymbol{r}}$ depending only on the initial position $\boldsymbol{r}$. As a result, the value function (1.14) can be identified with the mean first-passage time $\mathcal{T}(\boldsymbol{r})$ (MFPT) and the corresponding HJB equation (1.16) reads

$$(v_0\hat{\boldsymbol{u}}(\theta) + \boldsymbol{f}(\boldsymbol{r}))\cdot\nabla\mathcal{T} + D\nabla^2\mathcal{T} = -1\,, \tag{1.27}$$

where the gradients are taken with respect to $\boldsymbol{r}$. According to the optimality condition (1.17), the optimal choice for the heading direction θ (our control variable) is then obtained by

$$\partial_\theta[1 + (v_0\hat{\boldsymbol{u}} + \boldsymbol{f}(\boldsymbol{r}))\cdot\nabla\mathcal{T} + D\nabla^2\mathcal{T}] = 0 \Longrightarrow \tan\theta = \frac{\partial_y\mathcal{T}}{\partial_x\mathcal{T}} \iff \hat{\boldsymbol{u}}_\mathrm{opt}(\theta(\boldsymbol{r})) = \pm\frac{\nabla\mathcal{T}}{|\nabla\mathcal{T}|}\,, \tag{1.28}$$

valid at every point in space. As the desired strategy should minimize $\mathcal{T}(\boldsymbol{r})$, we expect the relevant control to point down its gradient and shall thus retain only the solution with a negative sign in (1.28). Inserting this into (1.27), we finally get the final expression for the stochastic HJB equation for the MFPT:

$$-\,v_0|\nabla\mathcal{T}| + \boldsymbol{f}(\boldsymbol{r})\cdot\nabla\mathcal{T} + D\nabla^2\mathcal{T} = -1\,, \tag{1.29}$$

to be solved with the boundary condition $\mathcal{T}(\boldsymbol{r}_\mathrm{T}) = 0$. The system composed by the last two Eqs. (1.28) and (1.29) hence provides the strategy which, by design, leads to the fastest possible trajectories on average. Hereafter, we will thus refer to it as the *Optimal Policy* (OP).

Figure 1.5b shows an example of the vector field corresponding to the OP control map $\hat{\boldsymbol{u}}_\mathrm{opt}(\theta(\boldsymbol{r}))$ obtained in the same setup used to illustrate the deterministic case (shown in Fig. 1.5a). Simulations of the Brownian dynamics (1.26) carried out with such control reveal that the stochastic trajectories tend to be symmetrically spread around Zermelo's path. This is consistent with the fact that Zermelo's solution is indeed recovered from OP in the limit $D \to 0$, an observation with important implications that will be discussed extensively in Chap. 4.

1.3.1 An Alternative Approach: Reinforcement Learning

Living organisms have shaped their behavior, functionality and morphology over millions of years in order to find the right balance between costs, risks and benefits. Evolution is therefore ultimately an optimisation process occurring as a result of the continuous interaction between an organism and its surroundings through what is called a *feedback loop*. This cycle, shown in Fig. 1.6, is also at the very heart of reinforcement learning (RL), a broad branch of machine learning algorithms that covers all possible decision-making problems [110], with potential applications ranging from engineering [111] and robotics [112, 113] to biology and active matter [23, 114–117].

In a nutshell, RL algorithms formalize the concept of learning how to carry out a task via trials and error. Let us imagine an *agent* moving in a certain *environment* (please refer to Fig. 1.6). Its actions lead to a change in the state of the environment (e.g. its position), which are then externally interpreted into rewards and finally fed back into the agent. Through a number of iterations, this agent tries to maximize the cumulative reward given a sequence of actions and, based on the experience gathered, it eventually learns the (ideally) optimal strategy.

In contrast to OC theory, approaches based on RL make it possible to study problems in which the dynamics of the environment is not exactly known a priori [118]. Let us imagine a glider that has to be airborne for as long as possible: it can only measure certain properties of its surroundings such as local temperature, pressure or wind direction, yet the pilot does not have enough information to be able to plan an

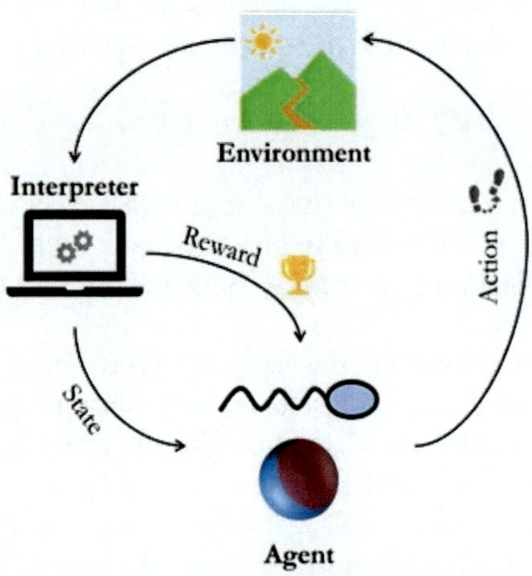

Fig. 1.6 Schematic representation of the feedback loop at the basis of both evolution of biological systems and RL algorithms. The agent, be it an artificial active particle, a robot or a living organism, modifies the state of the environment via its actions; it then receives updated information about the state of the environment together with a certain reward and, by gathering experience over time, will thus adapt its actions accordingly

optimal flight. However, a RL algorithm can, through experience in the field, infer the right manoeuvres to keep the glider in the air [111]. Furthermore, RL can help deal with problems in which the task is so complicated that it becomes extremely difficult, if not impossible, to determine the optimal strategy via OC.

This is why RL has recently been studied as an alternative numerical approach to tackle the problem of optimal navigation. These algorithms have indeed proven to be able to determine optimal strategies for navigating in complex environments [119–121] and even in the presence of chaotic [122] or turbulent flows [123–126]. Lastly, some RL-based methods have also shown the possibility of making smart microswimmers learn tactic behaviors [127–129] or how to escape from predators at low Reynolds number [130, 131].

1.3.2 Advantages and Drawbacks of Either Approach

Although a wide class of problems can be solved using either OC theory or RL algorithms, both these approaches suffer from inherent limitations.

First and foremost, the practical implementation of both requires the presence of an external interpreter either to guide (OC) or teach (RL) the agent how to move in an environment. Whilst such external control is today possible to perform on artificial microswimmers (e.g. via feedback loops [132, 133], hydrodynamic interactions with boundaries [46] or electromagnetic fields [134–136]), it severely limits the microswimmer autonomy and thus potential applications.

Besides, OC theory requires the knowledge of an accurate model of both the agent dynamics and the environment. This is in contrast with RL-based methods which are typically model-free, even though model-based approaches may provide several benefits like better asymptotic performance and faster convergence [137]. The application of OC strategies is also especially challenging to extend to more general situations such as active motion in a time-dependent turbulent flow [125] or on curved surfaces where solving the corresponding HJB equation (1.16) would require advanced computational techniques [138].

On the other hand, the navigation strategy obtained from OC theory is optimal by definition, which is something that is not guaranteed by the implementation of RL-based methods. The space of policies may in fact contain several local optima and convergence to the global optimum is not straightforward or even possible sometimes [114]. This is why the performance of RL algorithms should always be benchmarked against other known strategies [124, 125].

Moreover, the implementation of RL algorithms may demand a significant numerical effort. Indeed, it is often necessary to fine-tune the hyperparameters within the RL model in order to make the learning process converge while avoiding the risk of overfitting. This is to allow the learned strategies to be applicable also in environments slightly different than the one in which the learning took place [119].

Lastly, RL models with overly complex architectures, such as those based on deep neural networks [112, 113], despite allowing for the study of more sophisticated optimisation problems, have another intrinsic drawback: they act as black boxes that are so far from the physics of the problem as to make the interpretation of the resulting strategies rather difficult, with very few exceptions [127].

1.4 Current Challenges in Optimal Navigation: Thesis Outline

Although the problem of optimal navigation at the microscale has been extensively studied in recent years, be it via OC or RL, several challenges are still open in this context. On the one hand, for example, Zermelo's problem [108] has already been extended so to include energy dissipation and fuel consumption minimization [139] or the role of hydrodynamic interactions with obstacles [140]. On the other hand, however, we are still far from a treatment of the problem which is close enough to reality.

The aim of this work is therefore to extend the current state of the art in the field of optimal navigation in active matter systems. This will be achieved by introducing new theoretical tools, building bridges with existing techniques and designing new physics-informed heuristic models.

The content of this thesis is summarized in the diagram in Fig. 1.7.

In Chap. 2, we investigate the problem of optimal navigation for microswimmers moving on curved surfaces and in the presence of arbitrary stationary flows. There, we show that the solution can be mapped to finding the geodesics of an asymmetric Finsler-type metric known as Randers metric, thus providing a first link between microswimmer physics and the field of general relativity. We illustrate the application of this geometry-based strategy on both spherical and toroidal surfaces, showing that it always outperforms a more direct policy. The results contained in

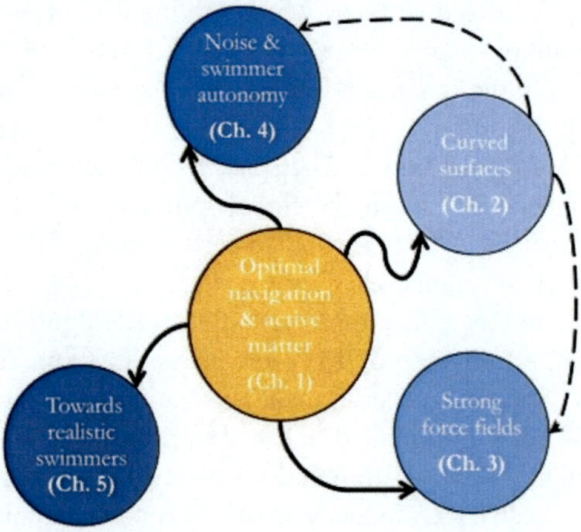

Fig. 1.7 Diagrammatic representation of the content of this thesis. The scope is to extend the state of the art in optimal navigation of active systems adding more and more ingredients to its description. The dashed arrows departing from Chap. 2 indicate that we have used the new theoretical tools introduced there also in Chaps. 3 and 4

this chapter have been published in https://journals.aps.org/prresearch/abstract/10.1103/PhysRevResearch.3.023125 Piro et al. *Phys. Rev. Research*, **3** 023125 (2021).

In Chap. 3, we study the performance and robustness of navigation strategies when motion takes place in the regime where the external force overcomes self-propulsion in finite regions. In particular, there we focus on the efficiency of Zermelo's classical solution in a periodic potential landscape. Despite allowing for a thorough exploration of the environment, our analysis indicates an increased sensitivity to initial conditions of the Zermelo strategy. This in turn suggests an interesting trade-off between exploration efficiency and stability for the design of control policies to be applied in actual systems. Lastly, we also investigate the same dynamical regime when in presence of finite space curvature by borrowing the theoretical tools introduced in Chap. 2. The content of this chapter has been adapted from https://www.frontiersin.org/articles/10.3389/fphy.2022.1034267/full Piro et al. *Front. Phys.*, **10** 1034267 (2022).

In Chap. 4, we include stochastic fluctuations in the description of our dynamical system. There, we address the problem of how microswimmers can optimize their navigation speed while not relying on the help of an external interpreter. To this end, we design new navigation strategies that can be implemented in a semi-autonomous fashion, as they do not require external real-time control over the microswimmer motion. Even though they rely on simple principles, these newly introduced strategies show performances strikingly similar to those obtained from SOC theory, proving also to be robust to environmental changes. Moreover, they have the advantage to be straightforwardly generalizable to even more complex optimization problems, like navigation on curved surfaces or even in presence of random flows. We finally show the existence of a new hierarchy of navigation protocols whose efficiency can be measured in terms of the enhancement of the effective drift along Zermelo's path. Part of the content of this chapter has been published in https://iopscience.iop.org/article/10.1088/1367-2630/ac9079 Piro et al. *New J. Phys.*, **24** 093037, (2022).

In Chap. 5, we finally add more ingredients to the model of our microswimmer so as to get closer to reality. In order to navigate, actual swimmers must in fact spend energy, and it is thus important to also consider their energetic efficiency in a thorough study of optimal navigation. To this end, we take into account varying self-propulsion speed and the swimmers' shape, which in turn indeed introduce novel interesting trade-offs between time and the energetic costs associated with active motion. Since swimmers with a well-defined shape can make use of the hydrodynamic torque exerted on their body to navigate, we investigate both the efficiency of *dumb* swimmers, i.e. whose motion is completely subject to the external flow, and *smart* swimmers who instead are capable of applying some control over their own dynamics. Our analysis thereby further extends Zermelo's classical solution and paves the way towards a comprehensive study of microswimmers' navigation efficiency.

References

1. Ramaswamy S (2017) Active matter. J Stat Mech: Theory Exp 2017(5):054002
2. Marchetti MC, Joanny JF, Ramaswamy S, Liverpool TB, Prost J, Rao M, Simha RA (2013) Hydrodynamics of soft active matter. Rev Mod Phys 85:1143–1189
3. Vicsek T, Zafeiris A (2012) Collective motion. Phys Rep 517(3):71–140. ISSN 0370-1573. https://doi.org/10.1016/j.physrep.2012.03.004, https://www.sciencedirect.com/science/article/pii/S0370157312000968
4. Chaté H (2020) Dry aligning dilute active matter. Annu Rev Condens Matter Phys 11(1):189–212
5. Whitesides GM, Grzybowski B (2002) Self-assembly at all scales. Science 295(5564):2418–2421
6. Carmona-Fontaine C, Deforet M, Akkari L, Thompson CB, Joyce JA, Xavier JB (2017) Metabolic origins of spatial organization in the tumor microenvironment. Proc Natl Acad Sci 114(11):2934–2939
7. Sweetlove LJ, Fernie AR (2018) The role of dynamic enzyme assemblies and substrate channelling in metabolic regulation. Nat Commun 9(1). https://doi.org/10.1038/s41467-018-04543-8
8. Agudo-Canalejo J, Golestanian R (2019) Active phase separation in mixtures of chemically interacting particles. Phys Rev Lett 123:018101
9. Cates ME, Tailleur J (2015) Motility-induced phase separation. Annu Rev Condens Matter Phys 6(1):219–244
10. Saha S, Agudo-Canalejo J, Golestanian R (2020) Scalar active mixtures: the nonreciprocal Cahn-Hilliard model. Phys Rev X 10:041009
11. Gompper G, Winkler RG, Speck T, Solon A, Nardini C, Peruani F, Löwen H, Golestanian R, Kaupp UB, Alvarez L, Kiørboe T, Lauga E, Poon WCK, DeSimone A, Muiños-Landin S, Fischer A, Söker NA, Cichos F, Kapral R, Gaspard P, Ripoll M, Sagues F, Doostmohammadi A, Yeomans JM, Aranson IS, Bechinger C, Stark H, Hemelrijk CK, Nedelec FJ, Sarkar T, Aryaksama T, Lacroix M, Duclos G, Yashunsky V, Silberzan P, Arroyo M, Kale S, (Feb2020) The, (2020) Motile active matter roadmap. J Phys: Condens Matter 32(19):193001
12. Berg HC (2008) E. coli in motion. Springer Science and Business Media
13. Kolomeisky AB, Fisher ME (2007) Molecular motors: a theorist's perspective. Annu Rev Phys Chem 58(1):675–695 PMID:17163836
14. Kassem S, van Leeuwen T, Lubbe AS, Wilson MR, Feringa BL, Leigh DA (2017) Artificial molecular motors. Chem Soc Rev 46:2592–2621
15. Bray D (2000) Cell movements. Garland science
16. Lin S-Z, Li Y, Ji J, Li B, Feng X-Q (2020) Collective dynamics of coherent motile cells on curved surfaces. Soft Matter 16:2941–2952
17. Berdahl A, Torney CJ, Ioannou CC, Faria JJ, Couzin ID (2013) Emergent sensing of complex environments by mobile animal groups. Science 339(6119):574–576
18. Cavagna A, Giardina I (2014) Bird flocks as condensed matter. Annu Rev Condens Matter Phys 5(1):183–207
19. Mahault B (2018) Outstanding problems in the statistical physics of active matter. Theses, Université Paris Saclay (COmUE). https://tel.archives-ouvertes.fr/tel-01880315
20. Needleman D, Dogic Z (2017) Active matter at the interface between materials science and cell biology. Nat Rev Mater 2(9). https://doi.org/10.1038/natrevmats.2017.48.
21. Akter M, Keya JJ, Kayano K, Kabir AMR, Inoue D, Hess H, Sada K, Kuzuya A, Asanuma H, Kakugo A (2022) Cooperative cargo transportation by a swarm of molecular machines. Sci Robot 7 (65):eabm0677. https://doi.org/10.1126/scirobotics.abm0677
22. Das S, Steager EB, Stebe KJ, Kumar V (2017) Simultaneous control of spherical microrobots using catalytic and magnetic actuation. In: 2017 international conference on manipulation, automation and robotics at small scales (MARSS), pp 1–6. https://doi.org/10.1109/MARSS.2017.8001924

23. Muiños-Landin S, Fischer A, Holubec V, Cichos F (2021) Reinforcement learning with artificial microswimmers. Sci Robot 6(52):eabd9285. https://doi.org/10.1126/scirobotics.abd9285
24. Dauchot O, Löwen H (2019) Chemical physics of active matter. J Chem Phys 151(11):114901
25. Popkin G (2016) The physics of life. Nature 529(7584):16–18. https://doi.org/10.1038/529016a
26. Shankar S, Souslov A, Bowick MJ, Marchetti MC, Vitelli V (2022) Topological active matter. Nature Reviews. Physics 4(6):380–398
27. Cavagna A, Del Castello L, Giardina I, Grigera T, Jelic A, Melillo S, Thierry Mora, Parisi L, Silvestri E, Viale M, Walczak AM (2015) Flocking and turning: a new model for self-organized collective motion. J Stat Phys 158:601–627. https://doi.org/10.1007/s10955-014-1119-3
28. Couzin DI, Krause J, James R, Ruxton GD, Franks NR (2002) Collective memory and spatial sorting in animal groups. J Theor Biol 218(1):1–11. ISSN 0022-5193. https://doi.org/10.1006/jtbi.2002.3065. https://www.sciencedirect.com/science/article/pii/S0022519302930651
29. Vicsek T, Czirók A, Ben-Jacob E, Cohen I, Shochet O (1995) Novel type of phase transition in a system of self-driven particles. Phys Rev Lett 75:1226–1229
30. Sumino Y, Nagai KH, Shitaka Y, Tanaka D, Yoshikawa K, Chaté H, Oiwa K (2012) Large-scale vortex lattice emerging from collectively moving microtubules. Nature. https://www.nature.com/articles/nature10874
31. Czirók A, Vicsek M, Vicsek T (1999) Collective motion of organisms in three dimensions. Phys A: Stat Mech Appl 264(1–2):299–304
32. Bechinger C, Di Leonardo R, Löwen H, Reichhardt C, Volpe G, Volpe G (2016) Active particles in complex and crowded environments. Rev Mod Phys 88:045006
33. Elgeti J, Winkler RG, Gompper G (Apr2015) Physics of microswimmers–single particle motion and collective behavior: a review. Rep Prog Phys 78(5):056601
34. Lauga E (2016) Bacterial hydrodynamics. Ann Rev Fluid Mech 48(1):105–130
35. Lauga E, Powers TR (2009) The hydrodynamics of swimming microorganisms. Rep Prog Phys 72(9):096601
36. Purcell EM (1977) Life at low reynolds number. Am J Phys45(3). https://doi.org/10.1119/1.10903
37. Berg HC, Anderson RA (1973) Bacteria swim by rotating their flagellar filaments. Nature 245(5425):380–382
38. Thompson MD, Mittelmeier TM, Dieckmann CL (2017) Chlamydomonas: the eyespot. Springer International Publishing, Cham, pp 257–281. ISBN 978-3-319-66365-4. https://doi.org/10.1007/978-3-319-66365-4_9
39. Wu SS, Kaiser D (1995) Genetic and functional evidence that type iv pili are required for social gliding motility in myxococcus xanthus. Mol Microbiol 18(3):547–558
40. Bershadsky AD, Kozlov MM (2011) Crawling cell locomotion revisited. Proc Natl Acad Sci 108(51):20275–20276
41. Chen Y-J, Nagamine Y, Yoshikawa K (2009) Self-propelled motion of a droplet induced by marangoni-driven spreading. Phys Rev E 80:016303
42. Marangoni C (1865) Sull'espansione delle goccie d'un liquido galleggianti sulla superfice di altro liquido. Fratelli Fusi. https://books.google.de/books?id=ZoxltwAACAAJ
43. Thutupalli S, Seemann R, Herminghaus S (2011) Swarming behavior of simple model squirmers. New J Phys 13(7):073021
44. Golestanian R, Liverpool TB, Ajdari A (2007) Designing phoretic micro- and nano-swimmers. New J Phys 9(5):126
45. Illien P, Golestanian R, Sen A (2017) 'fuelled' motion: phoretic motility and collective behaviour of active colloids. Chem Soc Rev 46(18):5508–5518
46. Das S, Garg A, Campbell AI, Howse J, Sen A, Velegol D, Golestanian R, Ebbens SJ (2015) Boundaries can steer active janus spheres. Nat Commun 6(1). https://doi.org/10.1038/ncomms9999
47. Golestanian R, Liverpool TB, Ajdari A (2005) Propulsion of a molecular machine by asymmetric distribution of reaction products. Phys Rev Lett 94:220801

48. Golestanian R (2018) Phoretic active matter. In: Active matter and nonequilibrium statistical physics, Lecture Notes of the Les Houches Summer School. Oxford University Press, London, England, November, p 2022
49. Xxvii Brown R (1828) A brief account of microscopical observations made in the months of June, July and August 1827, on the particles contained in the pollen of plants; and on the general existence of active molecules in organic and inorganic bodies. Philos Mag 4(21):161–173. https://doi.org/10.1080/14786442808674769
50. Einstein A (1905) Über die von der molekularkinetischen theorie der wärme geforderte bewegung von in ruhenden flüssigkeiten suspendierten teilchen. Annalen der Physik 322(8):549–560
51. Najafi A, Golestanian R (Jun2004) Simple swimmer at low reynolds number: three linked spheres. Phys Rev E 69:062901
52. Lauga E (2011) Life around the scallop theorem. Soft Matter 7:3060–3065
53. Volpe G, Buttinoni I, Vogt D, Kümmerer H-J, Bechinger C (2011) Microswimmers in patterned environments. Soft Matter 7:8810–8815
54. Callegari A, Volpe G (2019) Numerical simulations of active Brownian particles. Springer International Publishing, Cham, pp 211–238. ISBN 978-3-030-23370-9. https://doi.org/10.1007/978-3-030-23370-9_7
55. Khatami M, Wolff K, Pohl O, Ejtehadi MR, Stark H (2016) Active brownian particles and run-and-tumble particles separate inside a maze. Sci Rep 6:37670
56. Polin M, Tuval I, Drescher K, Gollub JP, Goldstein RE (2009) <i>chlamydomonas<i> swims with two "gears"; in a eukaryotic version of run-and-tumble locomotion. Science 325(5939):487–490
57. Howse JR, Jones RAL, Ryan AJ, Gough T, Vafabakhsh R, Golestanian R (2007) Self-motile colloidal particles: from directed propulsion to random walk. Phys Rev Lett 99:048102
58. Yanniotis S, Skaltsi S, Karaburnioti S (2006) Effect of moisture content on the viscosity of honey at different temperatures. J Food Eng 72(4): 372–377. ISSN 0260-8774. https://doi.org/10.1016/j.jfoodeng.2004.12.017. https://www.sciencedirect.com/science/article/pii/S0260877405000208
59. Segur JB, Oberstar HE (1951) Viscosity of glycerol and its aqueous solutions. Ind Eng Chem 43(9):2117–2120
60. James A, Plank MJ, Edwards AM (2011) Assessing Lévy walks as models of animal foraging. J Roy Soc Interface 8(62):1233–1247
61. Volpe G, Volpe G (2017) The topography of the environment alters the optimal search strategy for active particles. Proc Natl Acad Sci 114(43):11350–11355
62. Friedrich BM, Jülicher F (2007) Chemotaxis of sperm cells. Proc Natl Acad Sci 104(33):13256–13261
63. Vuijk HD, Merlitz H, Lang M, Sharma A, Sommer J-U (2021) Chemotaxis of cargo-carrying self-propelled particles. Phys Rev Lett 126:208102
64. Berg HC, Purcell EM (1977) Physics of chemoreception. Biophys J 20:193–219
65. Tsang ACH, Lam AT, Riedel-Kruse IH (2018) Polygonal motion and adaptable phototaxis via flagellar beat switching in the microswimmer euglena gracilis. Nat Phys 14(12):1216–1222
66. Engelmann ThW (1881) Neue methode zur untersuchung der sauerstoffausscheidung pflanzlicher und thierischer organismen. Pflüger, Archiv für die Gesammte Physiologie des Menschen und der Thiere 25(1):285–292. https://doi.org/10.1007/bf01661982
67. Dreyfus R, Baudry J, Roper ML, Fermigier M, Stone HA, Bibette J (2005) Microscopic artificial swimmers. Nature 437(7060):862–865
68. Huang H-W, Uslu FE, Katsamba P, Lauga E, Sakar MS, Nelson BJ (2019) Adaptive locomotion of artificial microswimmers. Sci Adv 5(1):eaau1532. https://doi.org/10.1126/sciadv.aau1532
69. Hong Y, Blackman NMK, Kopp ND, Sen A, Velegol D (2007) Chemotaxis of nonbiological colloidal rods. Phys Rev Lett 99:178103
70. Saha S, Golestanian R, Ramaswamy S (2014) Clusters, asters, and collective oscillations in chemotactic colloids. Phys Rev E 89:062316

71. Kranz WT, Gelimson A, Zhao K, Wong GCL, Golestanian R (2016) Effective dynamics of microorganisms that interact with their own trail. Phys Rev Lett 117:038101
72. Liebchen B, Löwen H (2018) Synthetic chemotaxis and collective behavior in active matter. Acc Chem Res 51(12):2982–2990
73. Jee A-Y, Dutta S, Cho Y-K, Tlusty T, Granick S (2018) Enzyme leaps fuel antichemotaxis. Proc Natl Acad Sci 115(1):14–18
74. Bennett RR, Golestanian R (2015) A steering mechanism for phototaxis in chlamydomonas. J Roy Soc Interface 12(104):20141164
75. Dai B, Wang J, Xiong Z, Zhan X, Dai W, Li C-C, Feng S-P, Tang J (2016) Programmable artificial phototactic microswimmer. Nat Nanotechnol 11(12):1087–1092
76. Lozano C, ten Hagen B, Löwen H, Bechinger C (2016) Phototaxis of synthetic microswimmers in optical landscapes. Nat Commun 7(1):12828
77. Kantsler V, Dunkel J, Blayney M, Goldstein RE (2014) Rheotaxis facilitates upstream navigation of mammalian sperm cells. eLife 3:e02403. ISSN 2050-084X. https://doi.org/10.7554/eLife.02403
78. ten Hagen B, Kümmel F, Wittkowski R, Takagi D, Löwen H, Bechinger C (2014) Gravitaxis of asymmetric self-propelled colloidal particles. Nat Commun 5(1):4829
79. Klumpp S, Faivre D (2016) Magnetotactic bacteria. Eur Phys J Special Topics 225(11):2173–2188
80. Matsunaga D, Meng F, Zöttl A, Golestanian R, Yeomans JM (2017) Focusing and sorting of ellipsoidal magnetic particles in microchannels. Phys Rev Lett 119:198002
81. Meng F, Matsunaga D, Golestanian R (2018) Clustering of magnetic swimmers in a poiseuille flow. Phys Rev Lett 120:188101
82. Kirkegaard JB, Goldstein RE (2018) The role of tumbling frequency and persistence in optimal run-and-tumble chemotaxis. IMA J Appl Math 83(4):700–719. ISSN 0272-4960. https://doi.org/10.1093/imamat/hxy013
83. Tsang ACH, Demir E, Ding Y, Pak OS (2020) Roads to smart artificial microswimmers. Adv Intell Syst 2(8):1900137. https://doi.org/10.1002/aisy.201900137
84. Sethi SP (2021) What is optimal control theory?. Springer International Publishing, Cham, pp 1–23. ISBN 978-3-030-91745-6. https://doi.org/10.1007/978-3-030-91745-6_1
85. Bellman R (1954) The theory of dynamic programming. Bull Am Math Soc 60(6):503–515. https://doi.org/10.1090/S0002-9904-1954-09848-8
86. Pontryagin LS (1987) Mathematical theory of optimal processes. Routledge. https://doi.org/10.1201/9780203749319
87. Bertsekas DP (2012) Dynamic programming and optimal control - 4th Edition, vol 2. Athena scientific Belmont, MA. http://www.athenasc.com/dpbook.html
88. Geering HP (2007) Optimal control with engineering applications. Springer. https://doi.org/10.1007/978-3-540-69438-0
89. Soner M (2005) Stochastic optimal control in finance. Edizioni della Normale
90. Bryson AE, Ho Y-C (2018) Applied optimal control: optimization, estimation, and control. Routledge
91. Kappen HJ (2005) Path integrals and symmetry breaking for optimal control theory. IOP Publishing 2005(11):11011
92. Risken H (1996) The Fokker-Planck Equation. Springer, Berlin Heidelberg, Berlin, Heidelberg
93. Peng S (1992) Stochastic Hamilton-Jacobi-bellman equations. SIAM J Control Optim 30(2):284–304
94. Yong J, Zhou XY (1999) Stochastic controls: Hamiltonian systems and HJB equations, vol 43. Springer Science & Business Media. https://doi.org/10.1007/978-1-4612-1466-3
95. Hein AM, Carrara F, Brumley DR, Stocker R, Levin SA (2016) Natural search algorithms as a bridge between organisms, evolution, and ecology. Proc Natl Acad Sci 113(34):9413–9420
96. Vidal-Mateo J, Benavent-Corai J, López-López P, García-Ripollés C, Mellone U, De la Puente J, Bermejo A, Urios V (2022) Search foraging strategies of migratory raptors under different environmental conditions. Front Ecol Evol 10:2022. ISSN 2296-701X. https://doi.org/10.3389/fevo.2022.666238

97. David C, Kennedy J, Ludlow A (1983) Finding of a sex pheromone source by gypsy moths released in the field. Nature 303:804–806
98. Tso W-W, Adler J (1974) Negative chemotaxis in escherichia coli. J Bacteriol 118(2):560–576. ISSN 0021-9193. https://jb.asm.org/content/118/2/560
99. Weihs D, Webb PW (1984) Optimal avoidance and evasion tactics in predator-prey interactions. J Theor Biol 106(2):189–206. ISSN 0022-5193. https://doi.org/10.1016/0022-5193(84)90019-5
100. Jiang J, Yang Z, Ferreira A, Zhang L (2022) Control and autonomy of microrobots: recent progress and perspective. Adv Intell Syst 4(5):2100279
101. Gao W, Wang J (2014) The environmental impact of micro/nanomachines: a review. ACS Nano 8(4):3170–3180 PMID: 24606218
102. Trincavelli M, Reggente M, Coradeschi S, Loutfi A, Ishida H, Lilienthal AJ (2008) Towards environmental monitoring with mobile robots. In: 2008 IEEE/RSJ international conference on intelligent robots and systems, pp 2210–2215. https://doi.org/10.1109/IROS.2008.4650755
103. Panda M, Das B, Subudhi B, Pati BB (2020) A comprehensive review of path planning algorithms for autonomous underwater vehicles. Int J Autom Comput 17(3):321–352. https://doi.org/10.1007/s11633-019-1204-9
104. Szczerba RJ, Galkowski P, Glicktein IS, Ternullo N (2000) Robust algorithm for real-time route planning. IEEE Trans Aerosp Electron Syst 36(3):869–878. https://doi.org/10.1109/7.869506
105. Chamseddine IM, Kokkolaras M (2018) Nanoparticle optimization for enhanced targeted anticancer drug delivery. J Biomech Eng 140(4):01. ISSN 0148-0731. https://doi.org/10.1115/1.4038202. 041002
106. Manzari MT, Shamay Y, Kiguchi H, Rosen N, Scaltriti M, Heller DA (2021) Targeted drug delivery strategies for precision medicines. Nat Rev Mater 6:351–370. https://doi.org/10.1038/s41578-020-00269-6
107. Park B-W, Zhuang J, Yasa O, Sitti M (2017) Multifunctional bacteria-driven microswimmers for targeted active drug delivery. ACS Nano 11(9):8910–8923
108. Zermelo E (1931) Über das navigationsproblem bei ruhender oder veränderlicher windverteilung. ZAMM J Appl Math Mech/Z Angew Math Mech 11(2):114–124. https://doi.org/10.1002/zamm.19310110205
109. Bonnans JF (2013) The shooting approach to optimal control problems. IFAC Proc Vol 46(11):281–292. ISSN 1474-6670. https://doi.org/10.3182/20130703-3-FR-4038.00158. https://www.sciencedirect.com/science/article/pii/S1474667016329597; 11th IFAC Workshop on Adaptation and Learning in Control and Signal Processing
110. Sutton RS, Barto AG (2018) Reinforcement learning: an introduction. MIT Press
111. Reddy G, Wong-Ng J, Celani A, Sejnowski TJ, Vergassola M (2018) Glider soaring via reinforcement learning in the field. Nature 562(7726):236–239. https://doi.org/10.1038/s41586-018-0533-0
112. Yang Y, Bevan MA, Li B (2020) Efficient navigation of colloidal robots in an unknown environment via deep reinforcement learning. Adv Intell Syst 2(1):1900106. https://doi.org/10.1002/aisy.201900106
113. Yang Y, Bevan MA, Li B (2020) Micro/nano motor navigation and localization via deep reinforcement learning. Adv Theory Simul 3(6):2000034
114. Cichos K, Gustavsson F, Mehlig B, Volpe G (2020) Machine learning for active matter. Nat Mach Intell 2:94–103. https://doi.org/10.1038/s42256-020-0146-9
115. Durve M, Peruani F, Celani A (2020) Learning to flock through reinforcement. Phys Rev E 102:012601
116. Falk MJ, Alizadehyazdi V, Jaeger H, Murugan A (2021) Learning to control active matter. Phys Rev Res 3:033291
117. Gerhard M, Jayaram A, Fischer A, Speck T (2021) Hunting active brownian particles: learning optimal behavior. Phys Rev E 104:054614
118. Powell WB (2019) From reinforcement learning to optimal control: a unified framework for sequential decisions. https://arxiv.org/abs/1912.03513

119. Monderkamp PA, Schwarzendahl FJ, Klatt MA, Löwen H (2022) Active particles using reinforcement learning to navigate in complex motility landscapes. https://arxiv.org/abs/2209.03047
120. Nasiri M, Liebchen B (2022) Reinforcement learning of optimal active particle navigation. New J Phys 24(7):073042
121. Schneider E, Stark H (2019) Optimal steering of a smart active particle. EPL (Eur Lett) 127(6):64003
122. Gustavsson K, Biferale L, Celani A, Colabrese S (2017) Finding efficient swimming strategies in a three-dimensional chaotic flow by reinforcement learning. Eur Phys J E 40(12). https://doi.org/10.1140/epje/i2017-11602-9.
123. Alageshan JK, Verma AK, Bec J, Pandit R (2020) Machine learning strategies for path-planning microswimmers in turbulent flows. Phys Rev E 101:043110
124. Biferale L, Bonaccorso F, Buzzicotti M, Clark Di Leoni P, Gustavsson K (2019) Zermelo's problem: optimal point-to-point navigation in 2d turbulent flows using reinforcement learning. Chaos 29:103138. https://doi.org/10.1063/1.5120370
125. Buzzicotti M, Biferale L, Bonaccorso F, Di Leoni PC, Gustavsson K (2021) Optimal control of point-to-point navigation in turbulent time-dependent flows using reinforcement learning. https://arxiv.org/abs/2103.00329
126. Colabrese S, Gustavsson K, Celani A, Biferale L (2017) Flow navigation by smart microswimmers via reinforcement learning. Phys Rev Lett 118:158004
127. Hartl B, Hübl M, Kahl G, Zöttl A (2021) Microswimmers learning chemotaxis with genetic algorithms. Proc Natl Acad Sci 118(19). ISSN 0027-8424. https://doi.org/10.1073/pnas.2019683118. https://www.pnas.org/content/118/19/e2019683118
128. Qiu J, Mousavi N, Gustavsson K, Xu C, Mehlig B, Zhao L (2022) Navigation of microswimmers in steady flow: the importance of symmetries. J Fluid Mech 932:A10. https://doi.org/10.1017/jfm.2021.978
129. Qiu J, Mousavi N, Zhao L, Gustavsson K (2022) Active gyrotactic stability of microswimmers using hydromechanical signals. Phys Rev Fluids 7:014311
130. Borra F, Biferale L, Cencini M, Celani A (2022) Reinforcement learning for pursuit and evasion of microswimmers at low reynolds number. Phys Rev Fluids 7:023103
131. Zhu G, Fang W-Z, Zhu L (2022) Optimizing low-reynolds-number predation via optimal control and reinforcement learning. J Fluid Mech 944:A3. https://doi.org/10.1017/jfm.2022.476
132. Fernandez-Rodriguez MA, Grillo F, Alvarez L, Rathlef M, Buttinoni I, Volpe G, Isa L (2020) Feedback-controlled active brownian colloids with space-dependent rotational dynamics. Nat Commun 11(1):4223
133. Khadka U, Holubec V, Yang H, Cichos F (2018) Active particles bound by information flows. Nat Commun 9:3864
134. Chen X-Z, Jang B, Ahmed D, Hu C, De Marco C, Hoop M, Mushtaq F, Nelson BJ, Pané S (2018) Small-scale machines driven by external power sources. Adv Mater 30(15):1705061. https://doi.org/10.1002/adma.201705061
135. Mano T, Delfau J-B, Iwasawa J, Sano M (2017) Optimal run-and-tumble based transportation of a janus particle with active steering. Proc Natl Acad Sci 114(13):E2580–E2589
136. Tierno P, Golestanian R, Pagonabarraga I, Sagues F (2008) Magnetically actuated colloidal microswimmers. J Phys Chem B 112(51):16525–16528
137. Moerland TM, Broekens J, Jonker CM (2020) Model-based reinforcement learning: a survey. CoRR, arXiv:2006.16712, https://arxiv.org/abs/2006.16712
138. Dziuk G, Elliott CM (2013) Finite element methods for surface pdes. Acta Numer 22:289–396. https://doi.org/10.1017/S0962492913000056
139. Liebchen B, Löwen H (2019) Optimal navigation strategies for active particles. EPL (Eur Lett) 127(3):34003
140. Daddi-Moussa-Ider A, Löwen H, Liebchen B (2021) Hydrodynamics can determine the optimal route for microswimmer navigation. Commun Phys 4(1):15

Chapter 2
Optimal Navigation Strategies for Active Particles on Curved Surfaces

Geometry is not true, it is advantageous.

Henri Poincaré

The content of this chapter has been adapted from Piro L., Tang E., Golestanian, R., *Phys. Rev. Research*, **3** 023125 (2021) [1]. Copyright ©2022 by the American Physical Society. I have taken part in the conception of the research and in the redaction of the manuscript. I have also designed the simulation code and analyzed the data.

Motion on curved surfaces and the role of local geometry have recently gained significant attention in the field of active matter [2] (e.g. in cells and tissues organization [3], cell motility [4] or collective motion of active particles [5, 6]). Here, we aim to develop an analytical formalism for optimal navigation in an over-damped system, which can be used on curved manifolds and arbitrary stationary flows.

Adopting recent mathematical results from differential geometry [7], we show that this problem can be mapped onto geodesics of a Finsler-type geometry with a Randers metric [8]. Finsler spaces are more general than Riemannian spaces since the tangent norms need not be induced by inner products [9]. They have been used to construct geometric descriptions in many areas of physics, with applications ranging from electron motion in magnetic fields [10] to quantum control [11] and test theories of relativity [12]. The particular choice of the asymmetric Randers metric moreover allows us to characterize the irreversibility of the optimal trajectory in this non-equilibrium problem.

We start by illustrating the formalism and discussing some general properties of the system. Then, we apply these concepts to some specific setups and study how following Randers geodesics can reduce the travel time to reach a target compared to when the microswimmer heads constantly towards it. Lastly, we analyze the *isochrones*—curves of equal travel time—to investigate more generally the shape of optimal paths.

L. Piro, *Optimal Navigation in Active Matter*, Springer Theses,
https://doi.org/10.1007/978-3-031-52577-3_2

2.1 Geomertry of Surfaces

Let us consider a two-dimensional smooth curved surface $\mathcal{M}$ embedded in a 3D Euclidean space $(\hat{\boldsymbol{x}}, \hat{\boldsymbol{y}}, \hat{\boldsymbol{z}})$, such that it can formally be defined as a function of two generic parameters q and p [13]

$$\boldsymbol{r}(q, p) = x(q, p)\hat{\boldsymbol{x}} + y(q, p)\hat{\boldsymbol{y}} + z(q, p)\hat{\boldsymbol{z}} \,.$$

As shown in Fig. 2.1, we can then identify a (not necessarily orthonormal) basis made up of two vectors lying in the locally tangent plane $T\mathcal{M}$, namely

$$\boldsymbol{t}_1(q, p) = \frac{\partial \boldsymbol{r}}{\partial q}\,, \quad \boldsymbol{t}_2(q, p) = \frac{\partial \boldsymbol{r}}{\partial p}\,, \tag{2.1}$$

together with the corresponding normal vector $\boldsymbol{n}$, defined as $\boldsymbol{n} = \boldsymbol{t}_1 \times \boldsymbol{t}_2$. Consequently, the square of the length of an infinitesimal displacement $\mathrm{d}\boldsymbol{r}$ along the surface can be written as

$$\mathrm{d}s^2 = \mathrm{d}\boldsymbol{r} \cdot \mathrm{d}\boldsymbol{r} = (\boldsymbol{t}_1 \mathrm{d}q + \boldsymbol{t}_2 \mathrm{d}p) \cdot (\boldsymbol{t}_1 \mathrm{d}q + \boldsymbol{t}_2 \mathrm{d}p) = h_{11}(\mathrm{d}q)^2 + 2h_{12}\mathrm{d}q\mathrm{d}p + h_{22}(\mathrm{d}p)^2 \,,$$

where we have just defined the components of the surface metric tensor[1]

$$h_{ij} \equiv \boldsymbol{t}_i \cdot \boldsymbol{t}_j \,, \text{with } \; i, j = 1, 2. \tag{2.2}$$

A surface equipped with this positive-definite tensor is known as a *Riemannian* manifold [15].

The metric tensor $\boldsymbol{h}$ allows to define distances and angles on $\mathcal{M}$. Indeed, the scalar product of any two tangent vectors $\boldsymbol{v}, \boldsymbol{w} \in T\mathcal{M}$ whose components are expressed in the Cartesian orthogonal basis $(\hat{\boldsymbol{x}}, \hat{\boldsymbol{y}}, \hat{\boldsymbol{z}})$ can be simply calculated via $\boldsymbol{v} \cdot \boldsymbol{w} \equiv h_{ij} v^i w^j$, where from now on summation over repeated indices is assumed. In the same way, the norm of any tangent vector $\boldsymbol{w} \in T\mathcal{M}$ is defined as $|\boldsymbol{w}|_h^2 = h_{ij} w^i w^j$. Lastly, hereafter the components of vectors which live in the tangent space are expressed using upper indices, while the lower index notation is defined by $w_i \equiv h_{ij} w^j$.

At this point, we should just provide a concrete version of the two parameters q and p. There is some freedom in choosing the parametrization that describes a surface. In some cases, its symmetries can help to identify a suitable choice. For instance, it is convenient to choose spherical coordinates to describe a sphere or the cylindrical ones for a cylinder and so on. More generally, a way to describe a surface in 3D is to introduce a constraint between x, y and z. The most straightforward example is the so-called *Monge representation* [13], where a surface is defined locally by its height $z(x, y)$ relative to the xy-plane, i.e.:

[1] In differential geometry, it is also known as the *first fundamental form* [14] of the manifold $\mathcal{M}$.

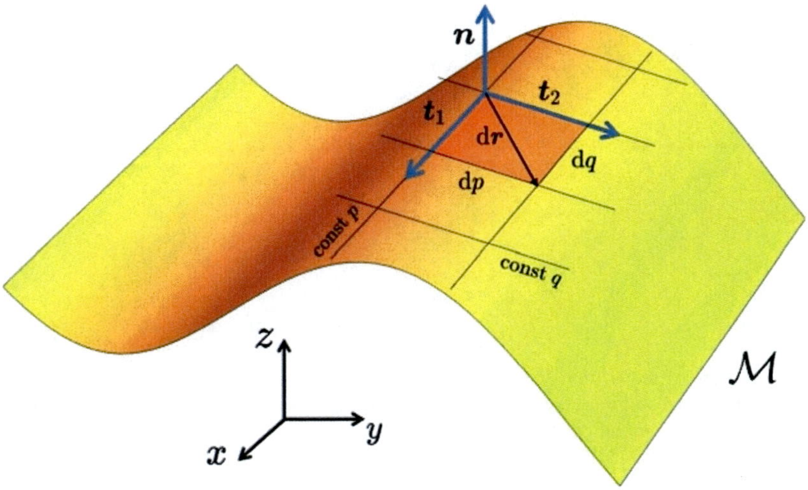

Fig. 2.1 Illustration of a generic surface $\mathcal{M}$ embedded in a three-dimensional Euclidean space. The Cartesian coordinates $(\boldsymbol{x}, \boldsymbol{y}, z)$ of the points on $\mathcal{M}$ depend on the two auxiliary variables q and p. These identify a local reference frame $(\boldsymbol{t}_1, \boldsymbol{t}_2, \boldsymbol{n})$ which allows to unambiguously characterize the surface

$$\boldsymbol{r}(x, y) = x\,\hat{\boldsymbol{x}} + y\,\hat{\boldsymbol{y}} + z(x, y)\,\hat{\boldsymbol{z}}\,. \tag{2.3}$$

The two tangent vectors area therefore given by

$$\boldsymbol{t}_1 = \hat{\boldsymbol{x}} + (\partial_x z)\hat{\boldsymbol{z}}, \qquad \boldsymbol{t}_2 = \hat{\boldsymbol{y}} + (\partial_y z)\hat{\boldsymbol{z}},$$

and the metric tensor $\boldsymbol{h}(x, y)$ is then straightforwardly obtained from (2.2):

$$\boldsymbol{h}(x, y) = \begin{pmatrix} 1 + (\partial_x z)^2 & (\partial_x z)\left(\partial_y z\right) \\ (\partial_x z)\left(\partial_y z\right) & 1 + \left(\partial_y z\right)^2 \end{pmatrix}. \tag{2.4}$$

2.2 Connection with Finsler Geometry

Let us now consider a microswimmer that is free to move at constant speed v_0 on a smooth Riemannian manifold $\mathcal{M}$ equipped with a positive-definite metric $\boldsymbol{h}$. The motion takes place in the presence of a time-independent force field $\boldsymbol{f}(\boldsymbol{r})$, which may in general include a contribution to due to advection by the solvent flow velocity (note that the friction coefficient is set to unity). The overdamped motion of the microswimmer can therefore be described as follows

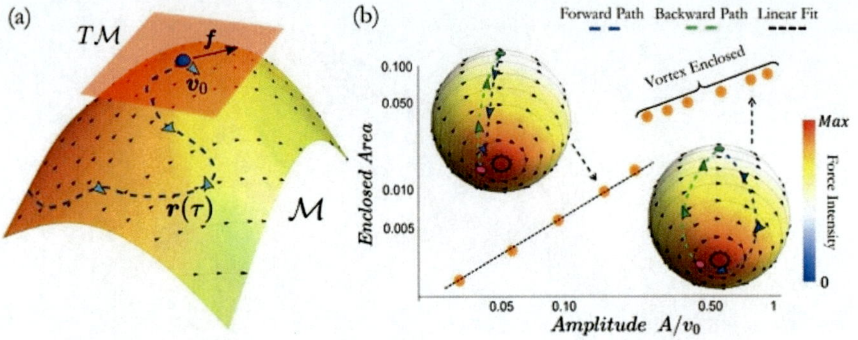

Fig. 2.2 **a** Microswimmer trajectory $\boldsymbol{r}(\tau)$ (blue dashed line) on a Riemannian manifold $\mathcal{M}$. The microswimmer moves in the tangent space $T\mathcal{M}$ under the joint influence of the force $\boldsymbol{f}$ (black arrows) and its self-propelling velocity $\boldsymbol{v}_0$ which is marked with cyan arrows. **b** Fraction of a spherical surface enclosed in the optimal forward-backward loop as a function of the force amplitude. The plot is in log-log scale. There is a jump when the vortex is enclosed in the loop. This introduces a qualitative difference which is reflected in the change from linear to sublinear scaling. There are two exemplar trajectories for given values of the force amplitude as indicated by dashed black arrows. The forward (blue dashing) and backward (green dashing) paths connect the following two points: $(\theta_0, \phi_0) = (\frac{\pi}{2}, 0) \rightleftarrows (\theta_1, \phi_1) = (\frac{11\pi}{12}, \frac{14\pi}{10})$. The arrows on top of each trajectory indicate the heading direction of the microswimmer. The intensity (color) gradient on each sphere shows the force intensity, while the solid black arrows show the force direction

$$\frac{\mathrm{d}\boldsymbol{r}}{\mathrm{d}\tau} = \boldsymbol{v}_0 + \boldsymbol{f}(\boldsymbol{r}(\tau)) \,, \tag{2.5}$$

where $\boldsymbol{v}_0 \equiv v_0 \hat{\boldsymbol{u}}$ corresponds to the swimmer self-propulsion velocity, while τ is the swimmer *proper* time (see Fig. 2.2a). We neglect rotational noise and assume full control over the direction $\hat{\boldsymbol{u}}$ of microswimmer propulsion. In practice, this means that the direction of propulsion is steered by a protocol that selects the appropriate active angular velocity to make it follow a prescribed path.

To show how Finsler geometry enters the optimal navigation problem on curved manifolds, we consider the time for a microswimmer to go from one point $\boldsymbol{r}_\mathrm{A}$ to another $\boldsymbol{r}_\mathrm{B}$ on the surface via the trajectory $\boldsymbol{r}(s)$ that is parametrized with s:

$$T = \int_{\boldsymbol{r}_\mathrm{A}}^{\boldsymbol{r}_\mathrm{B}} \mathrm{d}\tau = \int_{\boldsymbol{r}_\mathrm{A}}^{\boldsymbol{r}_\mathrm{B}} \frac{\mathrm{d}s}{v} \equiv \int_{\boldsymbol{r}_\mathrm{A}}^{\boldsymbol{r}_\mathrm{B}} \mathrm{d}s\, \mathcal{L}[s, \boldsymbol{r}(s), \dot{\boldsymbol{r}}(s)] \,, \tag{2.6}$$

where $v \equiv \frac{\mathrm{d}s}{\mathrm{d}\tau}$, $\dot{\boldsymbol{r}} \equiv \frac{\mathrm{d}\boldsymbol{r}}{\mathrm{d}s}$, and the Lagrangian $\mathcal{L} \equiv v^{-1}$ is defined by identifying the traveling time as an action. In order to find the explicit expression of $\mathcal{L}$, we can start from the equation of motion (2.5). Let us first isolate the self-propulsion term $v_0 \hat{\boldsymbol{u}}$ and then compute the squared norm of both sides of the equation, which leads to:

$$(h_{ij}\dot{r}^i\dot{r}^j)v^2 - (2h_{ij}\dot{r}^i f^j)v + f^2 - v_0^2 = 0 \,,$$

where we have used that $h_{ij}\hat{u}^i\hat{u}^j = 1$ and defined $f^2 \equiv h_{ij} f^i f^j$. Solving this quadratic equation for v, we easily obtain:

$$v = \frac{\pm\sqrt{h_{ij}\dot{r}^i\dot{r}^j(v_0^2 - f^2) + (h_{ij}\dot{r}^i f^j)^2} + h_{ij}\dot{r}^i f^j}{h_{ij}\dot{r}^i\dot{r}^j} \,. \tag{2.7}$$

Here, we now want to consider the most general case in which the intensity of the external force field can be larger than the self-propulsion. As opposed to the case where $f < v_0$, both signs in front of the square root satisfy the condition $v > 0$, ensuring that the particle never reverse its own motion along the path. However, since we looking for solutions that minimize the total travel time (2.6) we are clearly interested in finding the paths that maximize the total speed v of the active particle. As a result, we must consider the solution with the $+$ sign in (2.7). We thus now calculate the inverse of v and rationalize the denominator to arrive at an explicit expression for the Lagrangian:

$$\mathcal{L} = \frac{\sqrt{h_{ij}\dot{r}^i\dot{r}^j(v_0^2 - f^2) + (h_{ij}\dot{r}^i f^j)^2} - h_{ij}\dot{r}^i f^j}{v_0^2 - f^2} \,, \tag{2.8}$$

which can be further simplified by defining some new quantities: $\lambda^{-1} \equiv v_0^2 - f^2$, $a_{ij} \equiv h_{ij}\lambda + f_i f_j \lambda^2$, and $b_i \equiv -f_i\lambda$. This leads us to:

$$\mathcal{L} = \sqrt{a_{ij}\dot{r}^i\dot{r}^j} + b_i\dot{r}^i \,. \tag{2.9}$$

Since the expression (2.9) is positive definite, despite not being strongly convex,[2] it can still be identified as a Finsler metric [16, 17] of Randers type [8, 18, 19]. The only subtlety we need to consider is that the metric exhibits a singularity at $f = v_0$. However, by appropriately taking the limit $f \to v_0$, it is straightforward to show that the expression (2.8) reduces to

$$\mathcal{L} \underset{f\to v_0}{\longrightarrow} \frac{h_{ij}\dot{r}^i\dot{r}^j}{2h_{ij}\dot{r}^i f^j} \,, \tag{2.10}$$

which is still a well-defined positive quantity, also known in the mathematical literature as a Kropina metric [20].

[2] The strong-convexity condition is fulfilled if and only if $|f|_h < v_0$ at any point on the surface and any time [7].

2.3 Randers Spaces and Irreversibility

Randers spaces are often referred to as a special class of non-reversible Finsler spaces [18]. This is due to the presence of the second term in (2.9), namely $b_i\dot{r}^i$, which makes the metric tensor manifestly asymmetric under time reversal, i.e. $\mathcal{L}(\dot{r}^i) \neq \mathcal{L}(-\dot{r}^i)$. Due to this asymmetry, in presence of an external force the optimal forward path (between $\boldsymbol{r}_\mathrm{A}$ and $\boldsymbol{r}_\mathrm{B}$) will in general be different from the backward one ($\boldsymbol{r}_\mathrm{B}$ to $\boldsymbol{r}_\mathrm{A}$). In other words, the optimal backward path is distinct from the the time-reversed forward path, which highlights the out-of-equilibrium character of the navigation problem we study. In contrast, Riemannian geodesics (in the absence of any external force) are reversible since the corresponding metric tensor $\boldsymbol{h}$ is symmetric [21]. This property of Randers metrics is illustrated with a concrete example in Fig. 2.2b and studied in more details below.

We now exploit this geometrical analogy and solve the problem of optimal navigation on curved surfaces by calculating the Randers metric geodesic equation. To this end, since $\mathcal{L}$ is a homogeneous function of degree one with respect to $\dot{r}^i$, we can introduce the *fundamental tensor*

$$g_{ij} \equiv \frac{1}{2}\frac{\partial^2 \mathcal{L}^2}{\partial \dot{r}^i \partial \dot{r}^j}\,, \tag{2.11}$$

For the Randers metric (2.9), we then find

$$g_{ij} = \left(1 + \frac{b_i\dot{r}^i}{\sqrt{a_{ij}\dot{r}^i\dot{r}^j}}\right)\left(a_{ij} - \ell_i\ell_j\right) + \left(b_i + \ell_i\right)\left(b_j + \ell_j\right)\,, \tag{2.12}$$

where $\ell_i \equiv a_{ij}\dot{r}^j/\sqrt{a_{ij}\dot{r}^i\dot{r}^j}$. In order to determine the time-minimizing paths, we solve the Euler-Lagrange equations for the corresponding energy functional $\mathcal{E} = \frac{1}{2}\mathcal{L}^2 = g_{ij}\dot{r}^i\dot{r}^j$, namely $\frac{\mathrm{d}}{\mathrm{d}s}(\frac{\partial \mathcal{E}}{\partial \dot{r}^m}) = \frac{\partial \mathcal{E}}{\partial r^m}$. The paths minimizing $\int \mathrm{d}s\,\mathcal{E}$, which also minimize the travel time T, satisfy the Randers metric geodesic equation

$$\ddot{r}^k + \Gamma^k_{ij}\dot{r}^i\dot{r}^j = 0\,, \tag{2.13}$$

where the Christoffel symbol Γ^k_{ij} is defined via $\Gamma^k_{ij} \equiv \frac{1}{2}g^{km}(g_{im,j} + g_{jm,i} - g_{ij,m})$, with g^{km} being the inverse of the fundamental tensor defined in (2.11) and $g_{ij,m} \equiv \partial_m g_{ij}$.

Thus, the solutions of the geodesic equation (2.13) provide optimal navigation paths for a microswimmer moving in the presence of the force field $\boldsymbol{f}$ on a generic Riemannian manifold $\mathcal{M}$. Naturally, in flat spaces Randers geodesics correspond to the Zermelo trajectories introduced in Sect. 1.3.

It is important to note that the Christoffel symbols Γ^k_{ij} generally do not correspond to those associated with the metric $\boldsymbol{h}$, which are defined similarly as $\gamma^k_{ij} \equiv \frac{1}{2}h^{km}(h_{im,j} + h_{jm,i} - h_{ij,m})$. As natural geodesics correspond to shortest paths in absence of external force, the equality $\Gamma^k_{ij} = \gamma^k_{ij}$ is only satisfied if $f = 0$ everywhere.

2.4 Performance Assessment

We can now analyze the optimal paths obtained by following the Finsler geometry-based approach, which hereafter we call the *Randers Policy* (RP), in comparison with a benchmark, which we refer to as the *Straight Policy* (SP), in which the microswimmer always points in the direction of the target, regardless of the force field [22].

As a measure of performance, we compute the time T required to reach the target in units of the time T_R it would take in the absence of any external force, as a function of the maximum force on the surface. In the following of this chapter, we will focus on the regime in which the swimmer self-propulsion speed v_0 overpowers the external force $\boldsymbol{f}$ everywhere on the surface, namely $v_0 > \max |\boldsymbol{f}|_{\boldsymbol{h}}$.

2.4.1 *Optimal Navigation on a Sphere*

Let us first consider a sphere of radius unity embedded in $\mathbb{R}^3$. The position of the microswimmer on this surface can be written in spherical coordinates as $\boldsymbol{r} = (\theta, \phi)$. The corresponding Riemannian metric $\boldsymbol{h}$ in spherical coordinates has the components $h_{\theta\theta} = 1$, $h_{\phi\phi} = \sin^2\theta$, and $h_{\theta\phi} = h_{\phi\theta} = 0$. The force field is then $\boldsymbol{f}(\boldsymbol{r}) = f^\theta(\theta, \phi)\hat{\boldsymbol{e}}_\theta + f^\phi(\theta, \phi)\hat{\boldsymbol{e}}_\phi$.

2.4.1.1 Force Field with Two Vortices

As a first example, we choose $f^\theta(\theta, \phi) = 0$ and $f^\phi(\theta, \phi) = \frac{A\theta}{\pi \sin\theta}$, where A sets the amplitude of the field. This divergence-free force field is characterized by a pair of vortices at the poles of the sphere and its intensity is maximum (minimum) at the south (north) pole. We can then write the explicit expression of the Randers metrics $\mathcal{L}$ in our case as follows

$$\mathcal{L} = \frac{\sqrt{v_0^2 \sin^2\theta\dot{\phi}^2 + (v_0^2 - A^2\theta^2/\pi^2)\dot{\theta}^2} - A\dot{\phi}\theta\sin\theta/\pi}{v_0^2 - A^2\theta^2/\pi^2}.$$

It is then possible to determine the fundamental tensor g_{ij}, the relative Christoffel symbols Γ^k_{ij} and the corresponding geodesic equations using their definitions in (2.11) and (2.13). We further choose the following initial conditions: $\theta(0) = \theta_0$, $\phi(0) = \phi_0$, $\dot{\theta}(0) = -\sin\varphi_0 + f^\theta(\theta_0, \phi_0)$, and $\dot{\phi}(0) = \frac{\cos\varphi_0}{\sin\theta_0} + f^\phi(\theta_0, \phi_0)$. Here, (θ_0, ϕ_0) is the starting position while φ_0 represents the initial heading direction of the microswimmer (measured counterclockwise with respect to the $\hat{e}_\phi$ direction), which we scan when using the shooting method, selecting the one that takes the shortest time. Moreover, we parametrize the trajectory using the proper time of the microswimmer (i.e. we set $s = \tau$), which implies that $\mathcal{L}$ will be a conserved quantity along these paths. A clear advantage of using this parametrization is that it is independent of the choice of a coordinate system and does not impose any restrictions on the shape of the trajectories.

Before moving to a more detailed analysis of the performance of this new policy, we can directly compare the forward and backward paths. This can be quantified by showing how the area of the portion of sphere enclosed in the forward-backward loop varies with the intensity of the external force, A. In Fig. 2.2b, we show the results obtained for one choice of initial and final points. The area enclosed in the loop grows as the intensity of the force increases, which is expected since both paths deviate more from the Riemannian geodesic (the optimal path in the absence of external force). Interestingly, the enclosed area undergoes a jump when the vortex at the south pole is encircled, as beyond a certain threshold in the force intensity the microswimmer can exploit the vortex to reach the goal more quickly and this causes an abrupt change in the shape of the optimal forward path. The scaling with A is affected by this change, going from being essentially linear (black dashed line) to sublinear with an exponent ~ 0.6.

In Fig. 2.3 we show the results obtained for two different choices of the initial and final points. In either case, for small values of the force, the two strategies do not show substantial differences in terms of performance. However, for the example shown on the left in Fig. 2.3, two particular situations can be observed. For larger values of the force (yellow and green regions) RP (blue crosses) exploits the presence of the vortex at the south pole and at the same time the relative gain with respect to SP (red circles) grows. In fact, following the former strategy makes it possible for the microswimmer to take up to 40% less time to reach the target. Moreover, for sufficiently large values of the force intensity (green region), SP also includes the vortex. This slightly helps the swimmer, although just for a small range of values (see the local minimum in the green region). In addition, the relative gain following RP is substantial (up to about 20% in terms of arrival time) even when this strategy does not imply the exploitation of any specific force field structures (see the plot on the right in Fig. 2.3).

2.4.1.2 Force Field with a Sink and a Spiral

As a second example, let us consider a more complicated force field described by $f^\theta(\theta, \phi) = \frac{A}{\sqrt{\mu}}(\theta - \pi)$ and $f^\phi(\theta, \phi) = \frac{A}{\sqrt{\mu}}(\theta + \pi)$, where we have defined $\mu = \max_\theta[\sin^2\theta(\theta + \pi)^2 + (\theta - \pi)^2]$, so that A sets the force amplitude. This force field is characterized by a singular point (sink) at the north pole and a spiral with null intensity in its center at the south pole.

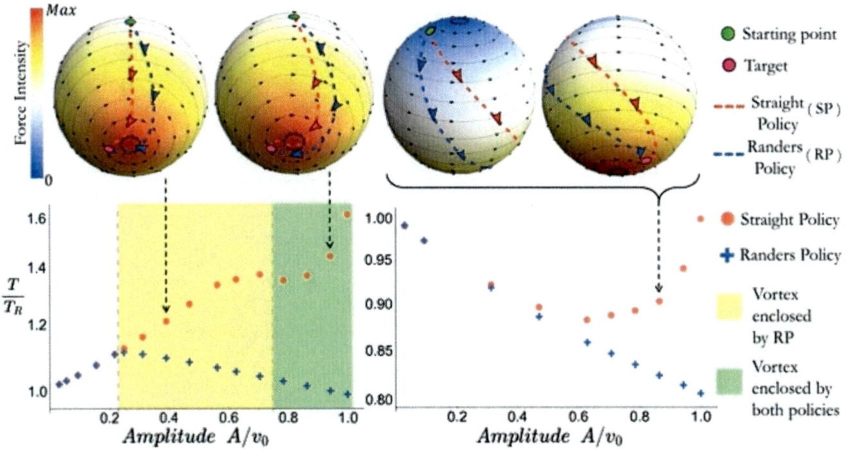

Fig. 2.3 Comparison between the Randers Policy (RP, blue crosses) and the Straight Policy (SP, red circles) in terms of the arrival times T in units of T_R—the optimal navigation time in the absence of f—as a function of the force amplitude. For both strategies, the target counts as reached when the microswimmer enters a region of radius $\epsilon = 0.01$. On the left: analysis of the paths connecting the points $(\theta_0, \phi_0) = (\frac{\pi}{2}, 0) \rightarrow (\theta_1, \phi_1) = (\frac{11\pi}{12}, \frac{14\pi}{10})$. On the right: study of the paths linking two diametrically opposite points $(\theta_0, \phi_0) = (\frac{\pi}{8}, 0) \rightarrow (\theta_1, \phi_1) = (\frac{7\pi}{8}, \pi)$. In either case SP is sub-optimal and there is a clear gain in following RP, especially when the force is stronger. In the upper part of the figure there are three different exemplar trajectories for given values of the force amplitude as indicated by the dashed black arrow. The dashed lines on every sphere represent the paths for each navigation strategy (RP: blue; SP: red) and the arrows on top of them show the corresponding heading direction of the microswimmer. The gradient on each sphere indicates the force intensity, while the solid black arrows represent its direction

The expression of the corresponding Randers metrics $\mathcal{L}$ (using its definition in (2.9)) is a bit more involved:

$$\mathcal{L} = \frac{\sqrt{(v_0^2 - \theta_{\pi_-}^2)s_\theta^2\dot{\phi}^2 + (v_0^2 - \theta_{\pi_+}^2 s_\theta^2)\dot{\theta}^2 + 2\theta_{\pi_+}\theta_{\pi_-}s_\theta^2\dot{\phi}\dot{\theta}} - (\theta_{\pi_+}s_\theta^2\dot{\phi} + \theta_{\pi_-}\dot{\theta})}{v_0^2 - (\theta_{\pi_+}^2 s_\theta^2 + \theta_{\pi_-}^2)},$$

where we have defined $\theta_{\pi_\pm} \equiv \frac{A}{\sqrt{\mu}}(\theta \pm \pi)$ and $s_\theta \equiv \sin\theta$. Then, once again we can determine the fundamental tensor g_{ij}, the Christoffel symbols Γ^k_{ij} and the corresponding geodesic equations simply using their definitions (2.11) and (2.13). Finally, we shall equip the obtained ODE system with the proper initial conditions and solve the navigation problem.

Notably, also with this choice of the force, we observe that there is always a clear gain in following the Randers Policy. The comparison with the performance achieved using the Straight Policy is shown in Fig. 2.4 where the displayed results correspond to two different choices of the target position.

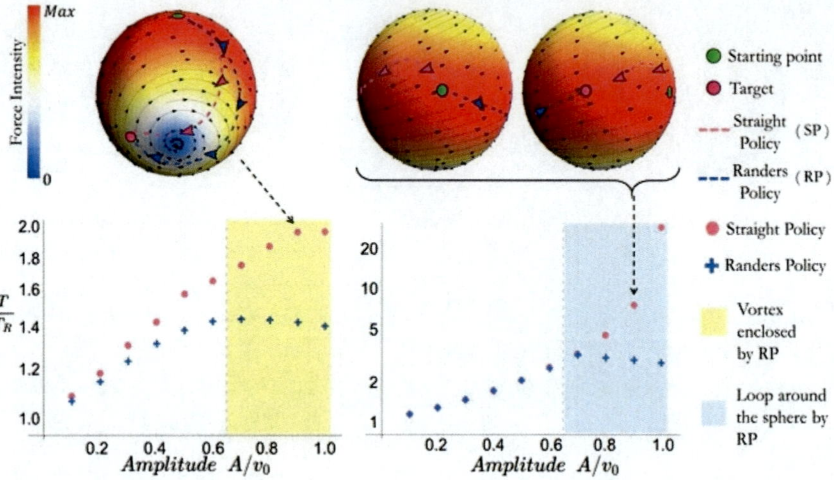

Fig. 2.4 Comparison between the arrival times of the Randers Policy (RP, blue crosses) and the Straight Policy (SP, magenta circles) as a function of the maximum force intensity on the sphere. On the left: analysis of the paths connecting the points $(\theta_0, \phi_0) = (\frac{\pi}{2}, 0) \rightarrow (\theta_1, \phi_1) = (\frac{5\pi}{6}, \frac{3\pi}{2})$. On the right: study of the paths linking the points $(\theta_0, \phi_0) = (\frac{\pi}{2}, 0) \rightarrow (\theta_1, \phi_1) = (\frac{\pi}{2}, \frac{5\pi}{3})$. In either case SP takes always longer than RP, which gets more advantageous the stronger the force. In the upper part of the figure there are two examples of how the trajectories look for a given value of the force as indicated by the dashed black arrow

Concerning the example on the left in Fig. 2.4, the exploitation of the spiral allows the swimmer following RP to take up to 30% less time to reach the target when compared to SP.

Remarkably, the relative gain achieved by following RP becomes extremely more significant (up to about 85% in terms of arrival time) when RP implies the swimmer to make a full turn around the sphere, as shown in the example on the right in Fig. 2.4 (blue region). This large difference between the two strategies is also due to the fact that following SP entails the swimmer having to point against the force direction where it is most intense (on the equator), making this policy highly disadvantageous.

2.4.2 *Optimal Navigation on a Torus*

Let us now show how one can use the same formalism to address the problem of optimal navigation in a different geometry. The motion takes place on a torus and the position of the microswimmer on its surface can be defined by:

Fig. 2.5 Scheme of the parametrization of a toroidal surface

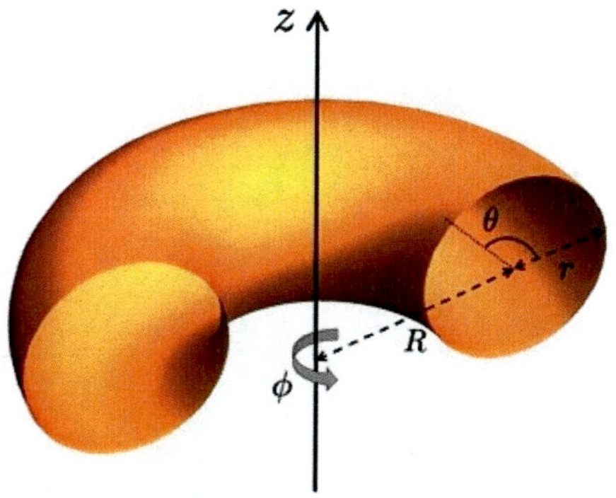

$$\begin{cases} x(\theta, \phi) = (R + r\cos\theta)\cos\phi \\ y(\theta, \phi) = (R + r\cos\theta)\sin\phi \\ z(\theta, \phi) = r\sin\theta, \end{cases}$$

in terms of the angles θ and ϕ, respectively the poloidal and toroidal directions (see Fig. 2.5). Let us fix the two characteristic radii of the torus to be $R = 2$ and $r = 1$ without loss of generality. The corresponding Riemannian metric h has the components $h_{\theta\theta} = 1$, $h_{\phi\phi} = (2 + \cos\theta)^2$ and $h_{\theta\phi} = h_{\phi\theta} = 0$. The force field is then $\boldsymbol{f}(\theta, \phi) = f^{\theta}(\theta, \phi)\hat{\boldsymbol{e}}_{\theta} + f^{\phi}(\theta, \phi)\hat{\boldsymbol{e}}_{\phi}$. For the sake of illustration, we consider a force field described by the following equations:

$$\begin{cases} f^{\theta}(\theta, \phi) = \frac{A}{\sqrt{2}}\sin\phi \\ f^{\phi}(\theta, \phi) = \frac{A}{\sqrt{2}}\frac{\cos\theta}{(2+\cos\theta)} \end{cases}, \tag{2.14}$$

where A sets its amplitude. This force field is characterized by a pair of vortices and another of saddle points with null intensity in their centers.

We can now write the expression of the Randers metrics $\mathcal{L}$ in this new case. Using the definitions in (2.9) and (2.14), we get to:

$$\mathcal{L} = \frac{\sqrt{c_{\theta_{+2}}^2(v_0^2 - \frac{A^2}{2}s_\theta^2)\dot{\phi}^2 + (v_0^2 - \frac{A^2}{2}c_\theta^2)\dot{\theta}^2 + A^2 s_\phi c_\theta c_{\theta_{+2}}\dot{\phi}\dot{\theta}} - \frac{A}{\sqrt{2}}(c_\theta c_{\theta_{+2}}\dot{\phi} + s_\phi\dot{\theta})}{v_0^2 - \frac{A^2}{2}(c_\theta^2 + s_\phi^2)},$$

where for simplicity we have defined $c_{\theta_{+2}} \equiv 2 + \cos\theta$, $c_\theta \equiv \cos\theta$, $s_\theta \equiv \sin\theta$ and $s_\phi \equiv \sin\phi$. As already done for the sphere, it is then possible to determine the fundamental tensor g_{ij}, the Christoffel symbols Γ^k_{ij} and the corresponding geodesic equations using their definitions reported in the previous section.

We further choose the following initial conditions: $\theta(0) = \theta_0$, $\phi(0) = \phi_0$, $\dot{\theta}(0) = \sin\varphi_0 + f^{\theta}(\theta_0, \phi_0)$, $\dot{\phi}(0) = \frac{\cos\varphi_0}{(2+\cos\theta_0)} + f^{\phi}(\theta_0, \phi_0)$, where the initial heading direction of the microswimmer φ_0 is measured counterclockwise with respect to the

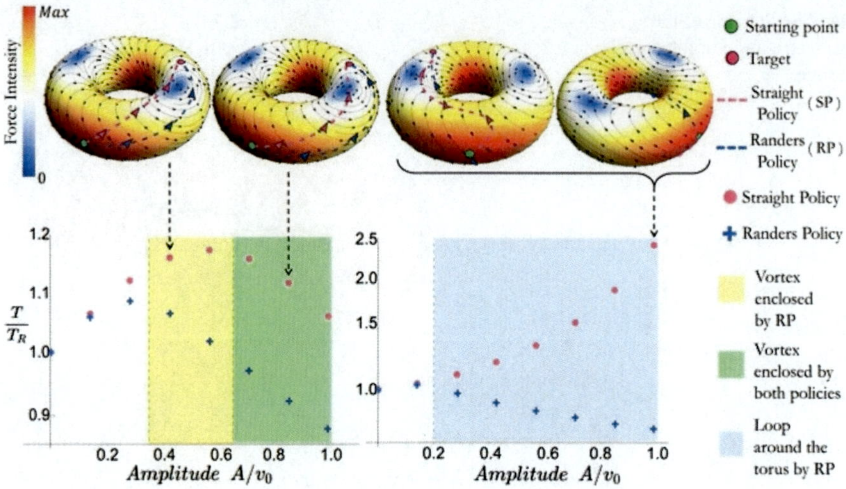

Fig. 2.6 Performance comparison between the Randers Policy (RP, blue crosses) and the Straight Policy (SP, magenta circles). For both strategies, the target counts as reached when the microswimmer enters a region of radius $\epsilon = 0.05$. On the left: analysis of the paths connecting the points $(\theta_0, \phi_0) = (0, \frac{3\pi}{2}) \rightarrow (\theta_1, \phi_1) = (\frac{\pi}{2}, \frac{\pi}{4})$. On the right: study of the paths linking the points $(\theta_0, \phi_0) = (0, \frac{3\pi}{2}) \rightarrow (\theta_1, \phi_1) = (\frac{\pi}{2}, \frac{3\pi}{4})$. In either case there is a clear gain in following RP, especially when the force is stronger. In the upper part of the figure there are three different examples of how the paths look for a given value of the force as indicated by the dashed black arrow. The dashed lines on every sphere represent the paths for each navigation strategy (RP: blue; SP: magenta) and the arrows on top of them show the corresponding heading direction of the microswimmer. The gradient on each sphere indicates the force intensity, while the solid black arrows represent its direction

$\hat{\phi}$ direction. Once again, we find the time-optimal paths by means of a shooting method and parametrize the microswimmer trajectories using its proper time, i.e. setting $s = \tau$.

We can now directly compare the performance achieved with the Randers Policy (RP) with that of the Straight Policy (SP). In Fig. 2.6 we show the analysis of the comparison between the arrival times obtained for two different choices of the initial and final points on the torus.

In both cases, the swimmer following RP has an increasing gain as the intensity of the force grows, confirming what has already been observed on the sphere. For the example shown on the left in Fig. 2.6, you can see how exploiting the presence of the vortex allows the swimmer which follows RP to reach the target faster (yellow and green regions). In fact, following this strategy makes it possible for the microswimmer to take up to 20% less time when compared to SP. Moreover, for sufficiently large values of the force intensity (green region), SP also includes the vortex, slightly helping the swimmer.

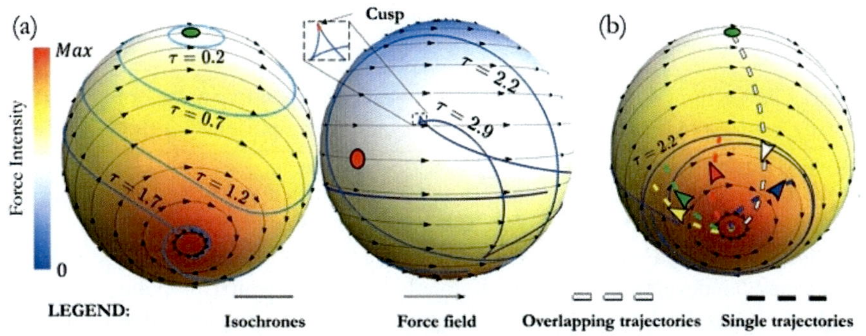

Fig. 2.7 Analysis of the isochrones starting from the point $(\theta_0, \phi_0) = (\pi/2, 0)$ (green circle) in the presence of the force field characterized by two vortices with $A = \frac{1}{2}v_0$. The red circle indicates the point diametrically opposite to the starting point and provides a guide to the eye. The color code on each sphere shows the force intensity from small (blue) to high (red), while the black arrows represent its direction. The time τ is reported in units of v_0^{-1}. **a** Isochrones (solid lines) at six different times. On the left: view of a region of stronger force. On the right: front view with an enlargement of a cusp highlighted in red, a point with a high density of geodesics. Also, notice the presence of self-intersections, points where two optimal paths collide. **b** Representation of 4 specific paths (dashed lines) passing close to the center of the vortex at the south pole. The arrows on top of them show the corresponding heading direction of the microswimmer and their color refers to the starting angle (green: $\varphi_0 = 4.7122$, blue: $\varphi_0 = 4.7123$, red: $\varphi_0 = 4.7124$, yellow: $\varphi_0 = 4.7125$). In the initial part where they overlap, the path is highlighted in white. This shows a strong dependence on initial conditions for optimal trajectories passing close to the vortex at the south pole

In addition, the relative gain achieved by following RP is even more significant (up to about 70% in terms of arrival time) when RP strategy implies the exploitation of the topological properties of the surface, as you can see from the example on the right in Fig. 2.6 (blue region).

2.5 Isochrone Analysis

To study more generally the behavior and the shape of the optimal trajectories coming from RP, we analyze the so-called *isochrones* in the illustrative examples of the sphere introduced in Sect. 2.4.1. These are curves of equal travel time obtained by fixing the microswimmer initial position (θ_0, ϕ_0) and varying the starting angle φ_0 from 0 to 2π. They can be seen as one-dimensional wavefronts of microswimmers that propagate onto the sphere following the Randers geodesics (2.13).

2.5.1 Force Field with Two Vortices

In Fig. 2.7a we show some isochrones (solid lines) corresponding to the optimal paths starting from a point on the equator (green circle), in the presence of the force field introduce in Sect. 2.4.1.1 with $A = \frac{1}{2}v_0$.

We observe that isochrones can feature self-intersections (see the example at $\tau = 2.9v_0^{-1}$ on the right in Fig. 2.7a). These are spots on the sphere for which there are multiple solutions to the problem of optimal navigation. At the same time, the isochrones can develop cusps, as highlighted in Fig. 2.7a, which are points at which neighboring geodesics meet. These cusp are analogues of *conjugate points* in general relativity [23], and related to the *caustics* in optics, as they represent domains on the isochrones with a higher density of geodesics [24]. Remarkably, these self-intersections and cusps occur only in the presence of a force field, as otherwise the isochrones on a sphere form concentric circles, and are a first signature of chaoticity in the system as we will show in Chap. 3.

Moreover, the isochrones are considerably distorted after they encounter the vortex at the south pole (see the isochrone at $\tau = 1.7v_0^{-1}$ in Fig. 2.7a). The meaning of such deformation can be understood by looking at Fig. 2.7b. Here are shown four optimal trajectories (dashed lines) starting from a point on the equator (green circle) and ending at time $\tau = 2.2v_0^{-1}$ on the corresponding isochrone (blue solid line). Their initial angles φ_0 differ only by $\Delta\varphi_0 = 10^{-4}$. Such paths initially overlap (white dashed line) and separate only once they reach the south pole. The observed strong dependence on initial conditions provides an interesting insight if one thinks of natural extensions of the model, e.g. the introduction of rotational noise. A direct implication of such abrupt non-linear effects is indeed that, in presence of a singularity in the force field, noise cannot be perturbatively included in the model and one must therefore consider alternative approaches [25, 26].

2.5.2 Force Field with a Sink and a Spiral

In Fig. 2.8 we show the isochrones (solid lines) corresponding to the optimal paths starting from a point on the equator (green circle), in the presence of the force field introduced in Sect. 2.4.1.2 with $A = 0.7v_0$.

Once again we find that there are spots on the sphere for which there are multiple solutions to the problem of optimal navigation, self-intersections, and others with a higher density of geodesics, cusps (see the example at $\tau = 3v_0^{-1}$ on the right in Fig. 2.8a).

However, here we can see a new peculiar feature of isochrones that has not been observed in the case illustrated above: they can be discontinuous (for e.g., see the isochrones at $\tau = 2.5v_0^{-1}$ and $\tau = 3v_0^{-1}$). This is certainly due to the presence of the singularity (sink) at the north pole and can be better understood by looking at Fig. 2.8b. Here are shown two optimal trajectories (dashed lines) starting from the

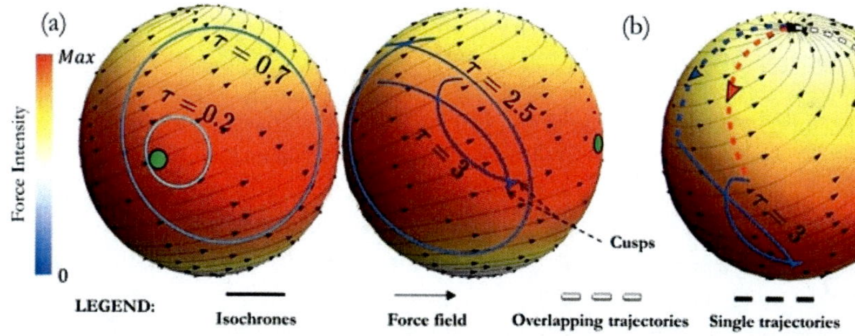

Fig. 2.8 Isochrones (solid blue lines) at four different times starting from the point $(\theta_0, \phi_0) = (\pi/2, 0)$ (green circle) in the presence of the force field introduced in Sect. 2.4.1.2 with $A = 0.7v_0$. **a** On the left: front view. On the right: side view, where the two isochrones display self-intersections, cusps and a discontinuity due to the presence of the singularity at the north pole. **b** Representation of two specific paths (dashed lines) passing close to the singularity at the north pole. The arrows on top of them show the corresponding heading direction of the microswimmer and their color refers to the starting angle (red: $\varphi_0 = 1.5707$, blue: $\varphi_0 = 1.5708$). In the initial part where they overlap, the path is highlighted in white

point on the equator and ending at time $\tau = 3v_0^{-1}$ on the corresponding isochrone (green solid line). Their initial angles φ_0 differ only by $\Delta\varphi_0 = 10^{-4}$. Such paths initially overlap (white dashed line) and separate only once they reach the north pole. Intuitively, since in its proximity there is an abrupt change in the force field direction, even a very small difference in the microswimmer position and orientation can lead to very different future scenarios. This prevents the use of perturbative methods to introduce noise into the model for trajectories passing in the neighbourhood of a singularity, as already pointed out above.

2.6 Summary and Discussion

In this chapter, we have formulated and discussed a geometric description of the optimal navigation problem for microswimmers on curved manifolds in the presence of a stationary force field. We have shown that this problem can be solved by finding the geodesics of a non-reversible Finsler metric of Randers type, providing a link between microswimmers physics and generalizations of general relativity. Notably, this formalism has the advantage of being sufficiently compact and elegant to enable the analysis of complex scenarios as well as possible generalizations.

Through the study of some illustrative cases, we have looked at how the area enclosed in the forward-backward loop varies as the force intensity increases, so as to show how the asymmetry of Randers metrics reflects in the out-of-equilibrium character of the optimal navigation problem. Then, comparing the performance of the Finsler geometry-based strategy (RP) with the one in which the swimmer always points towards the target (SP), we have found always a clear gain in following the former strategy. Moreover, such improvements get more significant as the force intensity grows.

Furthermore, the study of the shape of isochrones has revealed interesting general features concerning the optimal paths on curved surfaces. We have shown the presence of points with multiple solutions (self-intersections) as well as of points of higher density of geodesics (cusps). Thanks to the isochrones analysis we have also observed a strong dependence on initial conditions, especially for optimal trajectories passing close to the center of a vortex.

This raises the question of the optimal path stability which may severely limit the robustness of a strategy, especially when including noise into our model. As we will show in Chap. 3, isochrones can thus be a new powerful tool to measure the efficiency and stability of navigation policies in complex environments. At the same time, this novel approach may give us further insights about their performance in navigation tasks not requiring the presence of a specific target, such as when microswimmers try to escape from a harmful region [27].

Moreover, even though here we assumed to have full control over the microswimmer heading direction, our results could still be used to deal with positional and rotational diffusion. As will be discussed in more detail in Chap. 4, the microswimmer can indeed continuously adjust its orientation so as to stay close to the optimal path of the noise-free system [28].

Lastly, another assumption we made is that of stationary flow. Although this constraint cannot be relaxed as it is necessary to keep the analogy with Randers spaces and compute the corresponding geodesic equation, our results still hold true as long as there is a separation of timescales, with the characteristic timescale of the flow being much larger than the navigation time.

References

1. Piro L, Tang E, Golestanian R (2021) Optimal navigation strategies for microswimmers on curved manifolds. Phys Rev Res 3:023125
2. Castro-Villarreal P, Sevilla FJ (2018) Active motion on curved surfaces. Phys Rev E 97:052605
3. Callens SJP, Uyttendaele RJC, Fratila-Apachitei LE, Zadpoor AA (2020) Substrate curvature as a cue to guide spatiotemporal cell and tissue organization. Biomaterials 232:119739. ISSN 0142-9612. https://doi.org/10.1016/j.biomaterials.2019.119739. http://www.sciencedirect.com/science/article/pii/S0142961219308579
4. Lin S-Z, Li Y, Ji J, Li B, Feng X-Q (2020) Collective dynamics of coherent motile cells on curved surfaces. Soft Matter 16:2941–2952
5. Ehrig S, Ferracci J, Weinkamer R, Dunlop JWC (2017) Curvature-controlled defect dynamics in active systems. Phys Rev E 95:062609
6. Henkes S, Marchetti MC, Sknepnek R (2018) Dynamical patterns in nematic active matter on a sphere. Phys Rev E 97:042605
7. Bao D, Robles C, Shen Z (2004) Zermelo navigation on Riemannian manifolds. J Differ Geom 66(3):377–435
8. Randers G (1941) On an asymmetrical metric in the four-space of general relativity. Phys Rev 59:195–199

9. Finsler P (2018) Über kurven und flächen in allgemeinen räumen, Dissertation, University of Göttingen
10. Gibbons GW, Herdeiro CAR, Warnick CM, Werner MC (2009) Stationary metrics and optical zermelo-randers-finsler geometry. Phys Rev D 79:044022
11. Brody DC, Meier DM (2015) Solution to the quantum zermelo navigation problem. Phys Rev Lett 114:100502
12. Golestanian R, Khajehpour MRH, Mansouri R (1995) A test theory of the local structure of spacetime: a finslerian approach. Class Quant Gravity 12(1):273–278
13. do Carmo MP (1976) Differential geometry of curves and surfaces. Prentice Hall. ISBN 978-0-13-212589-5
14. Deserno M (2014) Notes on differential geometry. https://www.cmu.edu/biolphys/deserno/pdf/diff_geom.pdf
15. Schutz BF (1980) Geometrical methods of mathematical physics. Cambridge University Press
16. Javaloyes MA, Sánchez M (2017) Wind riemannian spaceforms and randers-kropina metrics of constant flag curvature. Eur J Math 3:1225–1244
17. Tang X, Yu C (2018) Some remarks on Einstein-Randers metrics. Differ Geom Appl 58: 83–102. ISSN 0926-2245. https://doi.org/10.1016/j.difgeo.2018.01.002. https://www.sciencedirect.com/science/article/pii/S0926224518300329
18. Bao D, Chern SS, Shen Z (2000) An introduction to Riemann-Finsler geometry. Springer, New York, NY
19. Cheng X, Shen Z (2012) Finsler geometry: an approach via Randers spaces. Springer
20. Yoshikawa R, Sabau SV (2014) Kropina metrics and zermelo navigation on riemannian manifolds. Geometriae Dedicata 171:119–148
21. Lee JM (1997) Riemannian geodesics. Springer New York, New York, NY, pp 65–89. ISBN 978-0-387-22726-9. https://doi.org/10.1007/0-387-22726-1_5. https://doi.org/10.1007/0-387-22726-1_5
22. Biferale L, Bonaccorso F, Buzzicotti M, Clark Di Leoni P, Gustavsson K (2019) Zermelo's problem: optimal point-to-point navigation in 2d turbulent flows using reinforcement learning. Chaos 29:103138. https://doi.org/10.1063/1.5120370
23. Manor Y (1977) Caustics in general relativity i. the phase function. Ann Phys 106(2):407–423. ISSN 0003-4916. https://doi.org/10.1016/0003-4916(77)90317-7. https://www.sciencedirect.com/science/article/pii/0003491677903177
24. Arnold V (1990) Singularities of caustics and wave fronts. Springer Netherlands. https://doi.org/10.1007/978-94-011-3330-2
25. Onsager L, Machlup S (1953) Fluctuations and irreversible processes. Phys Rev 91:1505–1512
26. Risken H (1996) The Fokker-Planck equation. Springer, Berlin, Heidelberg
27. Tso W-W, Adler J (1974) Negative chemotaxis in escherichia coli. J Bacteriol 118(2):560–576. ISSN 0021-9193. https://jb.asm.org/content/118/2/560
28. Piro L, Mahault B, Golestanian R (2022) Optimal navigation of microswimmers in complex and noisy environments. New J Phys 24(9):093037

Chapter 3
Efficiency of Navigation Strategies for Active Particles in Rugged Landscapes

Quelli che s'innamorano di pratica senza scienza son come il nocchiere, che entra in naviglio senza timone o bussola, che mai ha certezza dove si vada.

Leonardo da Vinci

The content of this chapter has been adapted from Piro L., Golestanian, R., Mahault B., *Front. Phys.*, **10** 1034267 (2022) [1]. I have taken part in the conception of the research and in the redaction of the manuscript. I have performed the numerical simulations and analyzed the corresponding data. I have also participated in the development of the simulation code.

Route optimization becomes especially important for active particles when the external influence of the environment is sufficiently strong to forbid certain routes or induce trapping. At the same time, the navigation strategies obtained by solving the Zermelo problem (discussed in Sect. 1.3) and its generalizations have been mostly theoretically studied for simple configurations and in the regime where the self-propulsion force is always greater than the external force [2–6] (see, however, Ref. [7]).

Here, we thus revisit the problem of optimal navigation considering self-propelled agents moving in two-dimensional potential landscapes and evaluate the ability of two different strategies to efficiently explore space in presence of a strong force field influencing the motion. Namely, we focus on the comparison between Zermelo's Policy (ZP, given by Eq. (1.21)) and the more trivial one in which the microswimmer picks an initial orientation and keeps it forever. This choice is what defines what we call the *Trivial Policy* (TP).[1]

To thoroughly study the efficiency of these strategies, we here make an extensive use of the isochrones, a new powerful tool already introduced in the previous chapter.

[1] This strategy is very similar, but not equivalent, to the Straight Policy (SP) introduced in Chap. 2. For SP, the microswimmer indeed does not maintain a constant bearing since it has to continuously adjust its orientation so as to point towards a specified target.

L. Piro, *Optimal Navigation in Active Matter*, Springer Theses,
https://doi.org/10.1007/978-3-031-52577-3_3

Indeed, given a distribution of initial orientations, these curves essentially delimit the potentially explored region up to some time and therefore provide quantitative information on how the optimal trajectories will manage to explore the surrounding space on average. Moreover, as already mentioned in Sect. 2.5, the analysis of the isochrones shape shall provide us with useful insights into the stability of a navigation strategy, which is key to know especially in the presence of fluctuations.

Performing extensive numerical simulations, our results reveal that, as expected, in a deterministic setting ZP systematically performs better than TP over all the available parameter space. The differences between the two policies are moreover particularly striking in the regime of large external force. Although in this case a significant proportion of trajectories following TP get arrested, thus strongly restricting the amount of space that can be visited, ZP manages to sustain ergodic exploration so long as there exist available routes. As our analysis shows, ZP's good performances follows from its ability to circumvent regions with a strong force field opposing the motion. On the other hand, we also show that in presence of strong external forces the dynamics produced by ZP is generally chaotic, which leads to an increased sensitivity to initial conditions and limits its long-time efficiency. When the dynamics is subject to fluctuations, ZP then becomes disadvantageous as it leads most of trajectories to be deflected towards trapping regions. In contrast, TP exhibits performance that are robust to the presence of fluctuations. Finally, considering a navigation setup on a curved sinusoidal surface, we show that these conclusions remain valid in presence of finite space curvature.

3.1 Optimal Navigation in a Confining Potential Landscape

We study the overdamped motion of a self-propelled particle moving on the plane at constant speed v_0 in presence of an externally applied stationary force field $\boldsymbol{f}$. The position $\boldsymbol{r} = x\hat{\boldsymbol{x}} + y\hat{\boldsymbol{y}}$ of the particle thus obeys Eq. (1.18), which we report here for convenience:

$$\dot{\boldsymbol{r}} = v_0\hat{\boldsymbol{u}}(\theta) + \boldsymbol{f}(\boldsymbol{r}), \tag{3.1}$$

where we implicitly set the particle motility to unity and $\hat{\boldsymbol{u}}(\theta)$ denotes the unit vector oriented along the direction defined by the angle θ. In order to study the efficiency of the navigation protocols under a possibly motion-limiting force field, we moreover consider a potential force $\boldsymbol{f}(\boldsymbol{r}) = -\nabla U(\boldsymbol{r})$ with

$$U(\boldsymbol{r}) = \frac{u}{4}\left[\sin(kx) + \sin(ky)\right] , \tag{3.2}$$

where u and $k = \pi/(2\ell)$ are parameters that set the strength and periodicity of the potential. The potential U consists of a square lattice of local minima and maxima, and the resulting force field is pictured in Fig. 3.1a. Rescaling space and time respectively with ℓ and ℓ/v_0, the dynamics (3.1) depends on a single control parameter $\gamma \equiv$

$\pi u/(8v_0)$ which measures the relative strength of the force with respect to self-propulsion.

The value of γ defines several dynamical regimes for the self-propelled motion. Indeed, for $\gamma < 1/\sqrt{2}$ the self-propulsion speed overcomes the strength of the external force over all the space, such that the particles can in principle travel everywhere without restriction. In the following we refer to this case as the *weak force regime*. On the contrary, the condition $\gamma > 1$ defines a *trapping regime* where the self-propelled particles are not able to escape local minima of the potential. Lastly, in the intermediate *confining regime* $1/\sqrt{2} < \gamma < 1$ the strength of the external drive is larger than self-propulsion only in disconnected regions between which the particles can travel almost freely.

To fully characterize the dynamics, we must specify an update rule for the self-propulsion orientation θ. The choice of such a rule amounts to selecting a specific navigation strategy, which in our case will be either TP, such that $\dot{\theta} = 0$, or by ZP as defined by Eq. (1.21), which we report again here for convenience:

$$\dot{\theta} = \sin^2\theta\, \partial_x f_y - \cos^2\theta\, \partial_y f_x + \sin\theta\cos\theta(\partial_x f_x - \partial_y f_y)\,. \tag{3.3}$$

All results presented below were obtained from numerical simulations performed by means of a fourth order Runge-Kutta method with a time step $\mathrm{d}t = 10^{-3}$. We checked that decreasing the value of $\mathrm{d}t$ did not qualitatively affect the results. For all policies, the isochrones were calculated simulating multiple trajectories with initial orientation θ_0 uniformly distributed on the circle. The corresponding angular resolution $\delta\theta_0$ was taken between 10^{-4} for TP and 10^{-6} for ZP. When specified, data fits were performed via the *FindFit* routine of Mathematica [8].

3.2 Isochrones Analysis on the Plane

In this section, we analyze the performances of ZP and TP for space exploration in the three dynamical regimes defined by the value of the parameter γ.

3.2.1 Short-Time Behavior of the Isochrones

Figure 3.1a shows typical isochrones obtained in 2D with both ZP and TP. They correspond to relatively short times as their overall surface extends to only a few periods of the potential. Starting from a local maximum of the potential (upper row in Fig. 3.1a), the isochrones are initially isotropic (pink curves) and at later times deform into a cross-like shape (blue curves) due to the local structure of the force field. In contrast, taking a saddle point as initial condition (lower row in Fig. 3.1a) the isochrones naturally elongate in the directions along which the force points away from the initial position. Although the shapes of the isochrones at later times resemble

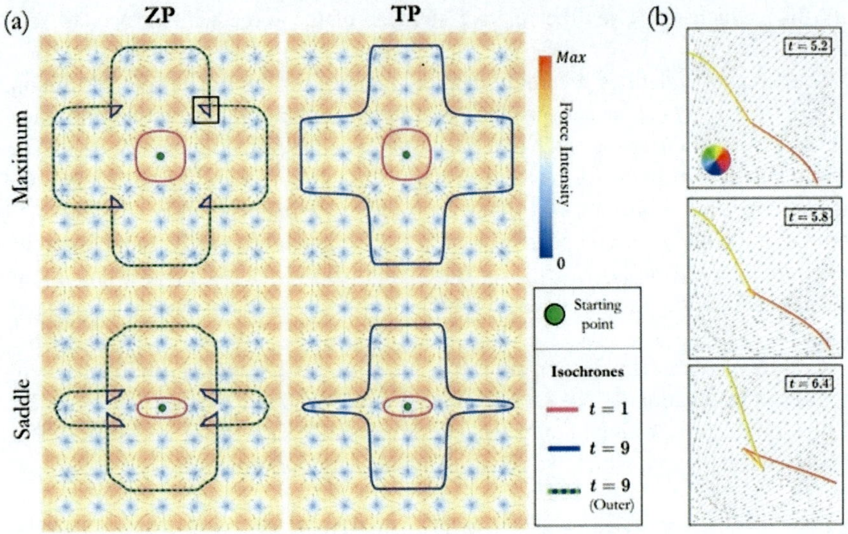

Fig. 3.1 Short-time isochrones in the confining regime. **a** Each panel shows two exemplary short-time isochrones at times $t = 1$ and $t = 9$ for a relative force strength $\gamma = 0.79$. The top and bottom rows correspond respectively to a departure point located at a local maximum or a saddle point of the potential (3.2), leading to different symmetries of the isochrones. Contrary to TP (right column), ZP (left column) exhibits characteristic self-crossings which allow trajectories to circumvent otherwise inaccessible regions of space. In all panels the colour map indicates the intensity of the external force field and the grey arrows indicate its direction. Panel **b** shows zooms of the region indicated by the black square in **b**. These three snapshots at different times show the formation of a self-intersection and two cusps in the ZP isochrone starting from a local maximum. The color code along the curve stands for the initial orientation of the trajectory as indicated by the color wheel

that of their counterpart originating from the local maximum, they still carry the signature of the initial anisotropy. Figure 3.1 moreover shows that, contrary to TP, the ZP isochrones develop self-crossing points in regions of strong and unfavorable force (i.e. pointing towards the departure position). As pictured in Fig. 3.1b, these self-crossings form because trajectories starting with neighboring initial orientations cross each other. As we will discuss below, these crossings are essential for ZP to be able to explore regions of space which would be unreachable otherwise.

The rest of this section is devoted to the characterization of the long-time isochrones properties.

3.2.2 *Isochrones Areas and Exploration Performances*

The first quantity of interest to measure the ability of the policies to efficiently explore the surrounding space is the isochrones area A. Figure 3.2a shows the scaling of A versus t^2. In absence of external drive ($\gamma = 0$), TP and ZP are strictly equivalent as the

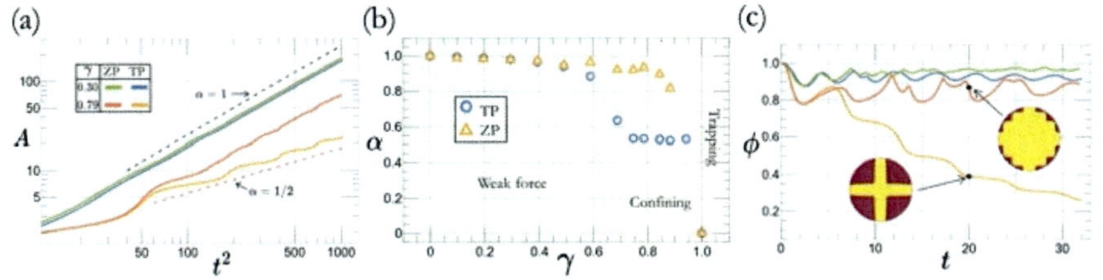

Fig. 3.2 The isochrones area analysis in the 2D periodic potential. **a** Isochrones area as function of squared time for ZP and TP and two values of the relative force strength γ respectively in the weak force and confining regimes. **b** Best fit of the exponent α (defined in Sect. 3.2) showing different scalings between TP and ZP in the confining regime. **c** Fractional area covered by the isochrones as function of time, the legend is the same as **a**. The insets show the outer boundary of the isochrones (yellow region) enclosed in the corresponding bounding disk (purple) for both policies at $t = 20$ and $\gamma = 0.79$

fastest way to travel between two points is to join them via a straight line. Therefore, in this limit isochrones expand isotropically along all directions and $A \sim t^2$ up to a constant prefactor. This behaviour moreover remains qualitatively valid throughout the weak force regime as the measured exponent α defined from the long-time scaling $A \sim t^{2\alpha}$ takes values close to 1 (Fig. 3.2b). To measure the isochrones anisotropy, we moreover define the *fractional area* ϕ as the ratio between the area A and that of the smallest disk containing the isochrone. For $\gamma = 0$, isochrones are perfect circles such that $\phi = 1$. Figure 3.2c shows that for $\gamma < 1/\sqrt{2}$ (in the weak force regime) ϕ oscillates due to the local force field structure, but takes a well-defined average value < 1. This value is systematically larger for ZP than for TP, indicating that isochrones associated to the former are more isotropic.

For $1/\sqrt{2} < \gamma < 1$ the dynamics (3.1)–(3.3) is in the confining regime, meaning that the amplitude of the force field overcomes that of the self-propulsion in certain regions of space. As a consequence, for TP the exponent α exhibits a sudden decrease from 1 to $\approx \frac{1}{2}$ at $\gamma \approx 1/\sqrt{2}$ (Fig. 3.2b). This behaviour is due to the fact that in this regime only trajectories starting with θ_0 close to a multiple of $\frac{\pi}{2}$ manage to progress away from the departure point after some time, while the others stop as the force created by the potential balances the self-propulsion. Consequently, the isochrones only grow along the horizontal and vertical axis, resulting in an increasingly strong anisotropy (see the inset of Fig. 3.2c). In this context, it is natural to expect that the isochrones area grows linearly in time, resulting in an exponent $\alpha = \frac{1}{2}$. This behavior is reminiscent of that of the initial spreading of the particle distribution in an infinite-horizon Lorentz gas [9]. These observations are moreover confirmed by the behaviour of the fractional area ϕ which is found to approach zero as $\phi \sim 1/t$. Hence, in confining environments TP generally allows to explore only a limited portion of the space.

For ZP, on the contrary, the situation is remarkably different. Indeed, in this case the area A keeps growing almost as t^2 such that the best fits of the exponents α provide values close to 1 even for $\gamma \gtrsim 0.8$ (see Fig. 3.2a, b). Similarly to the weak

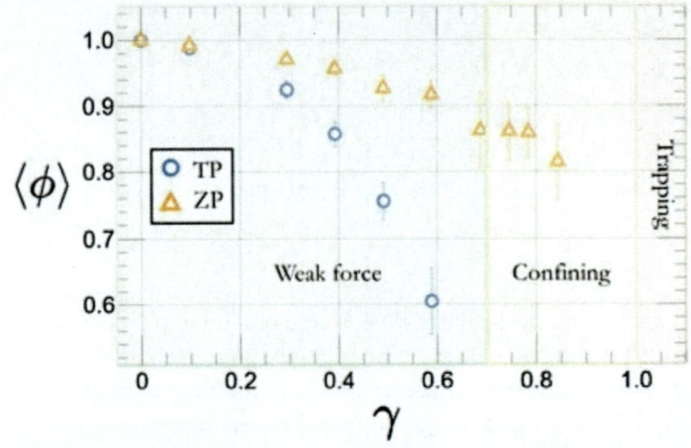

Fig. 3.3 Mean fractional area $\langle \phi \rangle$ covered by the isochrones at long times as function of the relative force strength γ. Both policies exhibit a decreasing trend in the weak force regime. ZP shows a plateau in the confining regime, while TP is not well-defined under such conditions (details in the text). Error bars correspond to the standard deviation due to oscillations of ϕ (see, e.g., Fig. 3.2c)

force regime, for ZP ϕ oscillates[2] around a well-defined value $\gtrsim 0.8$ even when the potential strongly constrains the dynamics (see Fig. 3.2c).

The fractional area ϕ covered by isochrones is a measure of how robust and efficient the strategy is at exploring the space isotropically. In the weak force regime, we observed that this quantity oscillates around a finite value for both ZP and TP. In such a case, it is possible to define a new performance indicator based on the mean value $\langle \phi \rangle$ around which they oscillate at long times. Figure 3.3 shows the comparison between the performance of the two policies in terms of $\langle \phi \rangle$ versus the relative force strength γ. Once again, ZP proves to be more efficient in exploring space, with $\langle \phi \rangle_{\text{ZP}}$ systematically larger than $\langle \phi \rangle_{\text{TP}}$.

It is moreover worth noting that $\langle \phi \rangle$ is well-defined in the confining regime only for ZP where it saturates above 0.8 (see the orange triangles in the yellow region of Fig. 3.3), showing that this strategy is still able to eventually cover the whole space even in presence of regions with an external force larger than self-propulsion. On the other hand, ϕ_{TP} does not oscillate around a finite value but rather decays to zero in the confining regime.

Therefore, and contrary to TP, the ZP isochrones manage to explore entirely the space despite the presence of stopping points in the force field (compare in particular the two insets of Fig. 3.2c). This remarkable feature however becomes harder to observe as γ approaches 1. Indeed, as we detail below for ZP the isochrones boundaries become sharper as γ grows, such that resolving them requires a rapidly increasing resolution. The limit $\gamma \to 1$ is thus numerically intractable but we expect the above conclusions to hold qualitatively until the trapping regime occurring at $\gamma = 1$ where the particles cannot travel further than a potential period regardless of the policy employed.

---

[2] The amplitude of fractional area oscillations moreover grows with γ as a result of the increasing influence of the potential landscape.

3.2.2.1 Isochrones Area Analysis Starting from a Different Point

In Sect. 3.2.1 we pointed out the isochrones anisotropy that arises at short times depending on the specific choice of the initial position $\boldsymbol{r}_0$, be it for instance a local maximum or a saddle point of the potential. However, as expected when the isochrones cover many periods of the potential, these shape differences do not qualitatively affect their long-time properties regardless of the chosen policy. In order to prove this, in Fig. 3.4a we compare the scaling of the isochrones area A as function of squared time t^2 in the confining regime. As can be observed, the curves corresponding to different starting points (maximum and saddle point of the 2D periodic potential) qualitatively show the same long-time scaling $A \sim t^{2\alpha}$ for both ZP ($\alpha = 1$) and TP ($\alpha = 1/2$).

A similar conclusion can be reached from Fig. 3.4b where we report the fractional area ϕ covered by the isochrones as function of time. After an initial transient where ϕ shows a sharper decrease when $\boldsymbol{r}_0$ is a saddle point rather than a maximum, the curves show similar trends at larger times. Convergence seems to be faster when starting from a local maximum than at a saddle point, which can simply be explained by the fact that the former case leads to faster trajectories (and thus spreading) initially.

Therefore, despite the quantitative shape differences observed at short times for different starting points, we have shown that the long-time properties of the isochrones remain qualitatively independent of the starting position. We therefore restrict the rest of the analysis to the case where the isochrones are initialized at a local maximum of the potential, corresponding to the upper row of Fig. 3.1a.

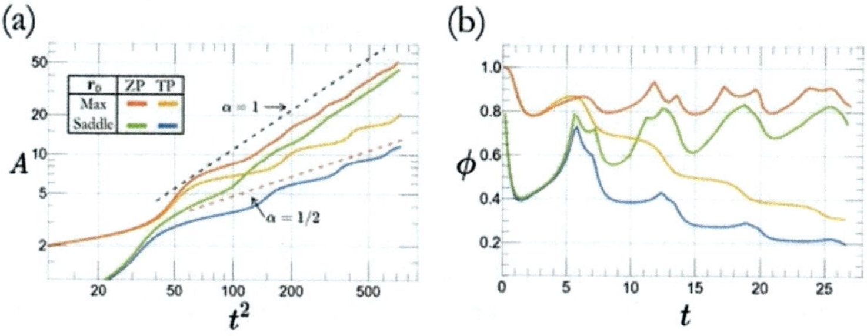

Fig. 3.4 Isochrones area analysis in the 2D periodic potential. **a** Isochrones area as function of squared time for ZP and TP and two different starting points, namely, a maximum and a saddle point of the force field whose relative strength here is fixed to $\gamma = 0.79$ (confining regime). **b** Fractional area covered by the isochrones as function of time, the legend is the same as **a**. This quantity oscillates around a finite value ~ 0.8 for ZP regardless of the starting point, while it goes to zero as $\phi \sim 1/t$ for the TP isochrones

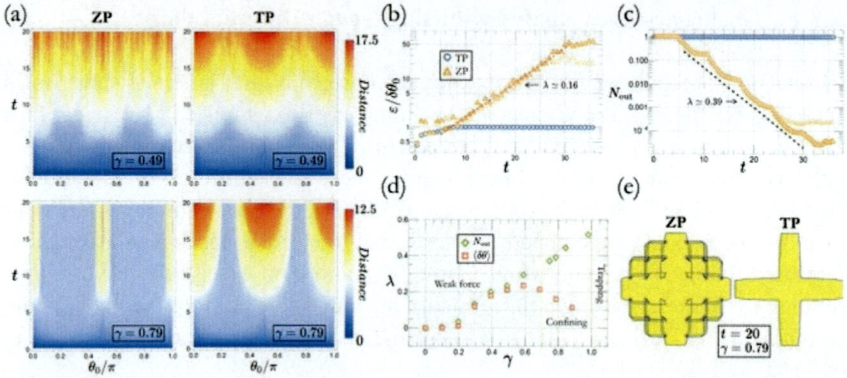

Fig. 3.5 Sensitivity of the navigation policies to the initial orientation. **a** Colour maps of the distance from the starting point as function of the initial angle θ_0 and time t for both ZP (left column) and TP (right column) in two different regimes: weak force ($\gamma = 0.49$, top row) and confining potential ($\gamma = 0.79$, bottom row). **b** Scaling with time of the mean angular separation ε defined in Eq. (3.4) for TP (blue circles) and ZP (orange triangles) at $\gamma = 0.79$. Opaque and semi-transparent symbols respectively correspond to an initial orientation resolution of $\delta\theta_0 = 10^{-6}$ and $\delta\theta_0 = 10^{-5}$. **c** The number of trajectories located on the outer isochrone as function of time, the legend and parameters are the same as **b**. **d** Best fits of the two sets of exponents obtained from the exponential scalings of ε and N_{out} (see **b** and **c**) for ZP as function of γ. **e** Example of long-time isochrones ($t = 20$) in the confining regime ($\gamma = 0.79$) for the two policies. In both cases the blue curve represents the entire isochrone, while the area highlighted in yellow is the region enclosed by the outer boundary. The complexity of the ZP isochrone is consistent with its larger sensitivity to the initial self-propulsion orientation

3.2.3 Robustness and Sensitivity to Initial Conditions

Interestingly, ZP manages to sustain ergodic exploration of space even in presence of a strongly confining force field. As we show now, this feat comes at the price of ZP being much more sensitive to the chosen initial orientation than TP.

Figure 3.5a shows colour maps of the distance reached by trajectories as function of their initial orientation θ_0 and time. In the weak force regime, isochrones grow nearly isotropically such that the distance reached by a trajectory after a certain time depends weakly on θ_0 (upper row of Fig. 3.5a). On the contrary, in the confining regime the potential landscape draws *escape routes*—corresponding for the parameters considered here to θ_0 being a multiple of π—along which the self-propelled particles travel significantly faster (lower row of Fig. 3.5a). Surprisingly, although ZP is globally better at exploring space, its escape routes are substantially narrower than that of TP. Therefore, considering a uniform distribution of initial orientations ZP counter-intuitively leads to a lower proportion of trajectories that manage to reach a given distance from the departure point than TP.

We now quantify the sensitivity to the initial orientation with the mean spatial angular separation

$$\varepsilon(t) \equiv \left\langle \arccos\left[\frac{\boldsymbol{r}(t|\theta_0)\cdot\boldsymbol{r}(t|\theta_0+\delta\theta_0)}{r(t|\theta_0)r(t|\theta_0+\delta\theta_0)}\right]\right\rangle_{\theta_0}, \tag{3.4}$$

where $\boldsymbol{r}(t|\theta_0)$ denotes the position at time t of the particle given the initial orientation θ_0 (taking the departure point for origin) and $r(t|\theta_0)$ is the associated distance, while $\delta\theta_0$ in Eq. (3.4) corresponds to the angular resolution of θ_0. Figure 3.5b, d show that for $\gamma \gtrsim 0.2$ the mean separation $\varepsilon(t)$ grows exponentially at large t for ZP over a time-range that increases as $\delta\theta_0 \to 0$. In contrast, for TP $\varepsilon(t)$ converges to a finite value, such that over long times the isochrones take a scale invariant form. Based on these observations, we may define a generalized Lyapunov exponent $\lambda_\varepsilon \equiv \ln[\varepsilon(t)]/t$ from the long time scaling of $\varepsilon(t)$. While $\lambda_\varepsilon^{\mathrm{TP}} = 0$ over the whole available range of γ, $\lambda_\varepsilon^{\mathrm{ZP}} > 0$ for $\gamma \gtrsim 0.2$, indicating that for sufficiently strong forces the dynamical system formed by Eqs. (3.1)–(3.3) is chaotic. Similar chaotic behavior of ZP was actually reported for navigation in turbulent flows [7].

We now build further insight from the analysis of *outer isochrones*, which we define as the minimal set of points that draw the boundary of the region enclosed by the isochrones (see e.g. the dashed green lines in Fig. 3.1a). We show in Fig. 3.5c that the number of trajectories $N_{\mathrm{out}}(t)$ remaining on the outer envelope of the isochrones is found to decay exponentially fast in time for ZP while stays constant for TP. The exponential decay of $N_{\mathrm{out}}(t)$ for ZP is explained by the increasing number of self-crossings of the isochrones with time (see Fig. 3.5e) which lead a large part of the trajectories to quit the boundary. We thus define from the long-time regime a second exponent $\lambda_N \equiv -\ln[N_{\mathrm{out}}(t)]/t$ such that, in agreement with the analysis of the angular separation ε, $\lambda_N^{\mathrm{TP}} = 0$ for all values of γ while $\lambda_N^{\mathrm{ZP}} > 0$ for γ large enough. As shown in Fig. 3.5d, $\lambda_\varepsilon^{\mathrm{ZP}} \approx \lambda_N^{\mathrm{ZP}}$ in the weak force regime but the exponents depart from each other when approaching the confining regime. Namely, although $\lambda_\varepsilon^{\mathrm{ZP}}$ is maximal at $\gamma \approx 0.6$ and starts to decay when entering the confining regime, λ_N^{ZP} increases almost linearly with γ up to the trapping regime at $\gamma = 1$. This difference of behaviours is well understood from the fact that the average in Eq. (3.4) is taken over all trajectories, including those that remain inside of the envelope formed by the outer isochrone, while $N_{\mathrm{out}}(t)$ only gets contributions from the most quickly diverging trajectories.

The multiple crossings observed on the ZP isochrones allow the trajectories to circumvent regions which are inaccessible when approached by facing the external force. Therefore, so long as $\gamma < 1$ ZP shall in principle ensure total coverage of space despite the presence of strong external force, in contrast with TP for which space exploration becomes quickly limited. However, due to the exponential decrease of the number of trajectories at the isochrones boundaries with time, the maximum area that can be spanned by ZP grows with the total number of trajectories N as $A_{\max} \sim \ln^2 N$, which limits the numerical exploration of the long-time regime as γ approaches 1. In practical situations, these results moreover suggest that for strong forces ZP is less reliable than TP, as its sensitivity to initial conditions can easily lead trajectories to depart from the predetermined route, which we now illustrate by including noise in the dynamics.

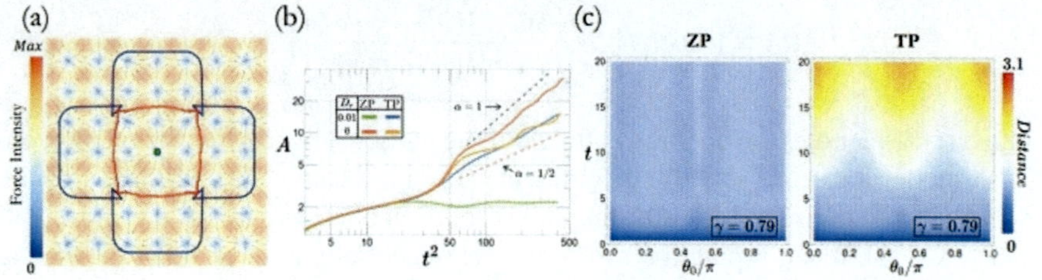

Fig. 3.6 **a** Comparison between the noiseless (blue curve) and noise-averaged (red curve) isochrones at time $t = 9$. Colour gradient indicates the intensity of the external force field and the grey arrows its direction. **b** Area enclosed by the isochrones as function of squared time for both ZP and TP with and without rotational noise. **c** Color maps of the average distance from the starting point as function of the initial orientation θ_0 and time t for both ZP (left column) and TP (right column) in presence of rotational noise. All data shown in **a–c** are averaged over 10^2 independent trajectories with a resolution $\delta\theta_0 = 10^{-2}$. The relative force strength is here set to $\gamma = 0.79$ (confining regime) and the rotational diffusion coefficient is equal to $D_r = 0.01$

3.2.4 *Effect of Fluctuations on the Navigation Performances*

The increased sensitivity of ZP to the initial particle orientation described previously raises the question of the impact of noise on the exploration performances. Fluctuations are a particularly important feature of the motion of swimmers evolving at the microscopic scale. As we have already discussed in Sect. 1.1.2, they can indeed arise due to thermal noise, but also due to fluctuations in the processes generating self-propulsion.

Here, we model the microswimmer as an ABP (see Eq. (1.2)) and further allow it to actively steer, such that its position $\boldsymbol{r}$ and self-propulsion orientation angle θ obey

$$\begin{cases} \dot{\boldsymbol{r}} = v_0\hat{\boldsymbol{u}}(\theta) + \boldsymbol{f}(\boldsymbol{r}) + \sqrt{2D}\boldsymbol{\xi} \\ \dot{\theta} = M(\boldsymbol{r}, \theta) + \sqrt{2D_r}\xi_\theta\ , \end{cases} \tag{3.5}$$

where D and D_r stand for the translational and rotational diffusivities respectively, while $\boldsymbol{\xi}$ and ξ_θ are independent delta-correlated white noises with unit variance. Finally, $M(\boldsymbol{r}, \theta)$ represents the active torque applied by the particle and depends on the chosen policy.[3] Our simulations moreover indicate that both sources of noise (translational and rotational) have a similar effect on the dynamics, such that we focus here on rotational noise, i.e. setting $D = 0$ hereafter.

Considering the weak force regime, we find that the performances of both policies do not change significantly with respect to the noiseless case. Conversely, in the confining regime the presence of rotational noise strongly impairs the ZP ability to explore the surroundings. This feature appears clearly from the behavior of the noise-averaged isochrone (Fig. 3.6a) whose area stops growing as soon as trajectories reach the local minima of the potential (see the green curve in Fig. 3.6b).

[3] For ZP, it is equal to the rhs of Eq. (3.3), while $M(\boldsymbol{r}, \theta) = 0$ for TP.

We thus conclude that for γ large enough the ZP escape routes identified in Fig. 3.5a are unstable, while the potential local minima act as attractors of the dynamics. Trajectories initially on the escape routes therefore deviate (on average) because of fluctuations and end up at the potential minima where they get stuck. This effect is moreover enhanced by the chaoticity of the ZP dynamics characterized previously, such that we expect it to be present for arbitrary weak noise. As a consequence, Fig. 3.6c shows that the mean distance reached by ZP trajectories is bounded for all initial orientations, leading to a disappearance of the escape routes. In contrast, we find that the properties of the TP isochrones are barely affected by noise, proving it to be a more robust strategy than ZP in presence of strong external forces.

3.3 Navigation on Curved Surfaces

In a number of situations, self-navigating agents move on non-Euclidean spaces including planes above the Earth [10, 11] or migrating cells [12, 13]. As the presence of a nonzero curvature locally leads to a stretching or compressing of distances, one naturally expects it to affect optimal navigation strategies.

In this section, building on the results derived in Chap. 2 we consider the optimal navigation problem on a two-dimensional curved surface in presence of strong external forces.

In contrast with the planar case, following natural geodesics on curved surfaces in presence of an external force requires for the self-propelled agent to adjust its direction of motion $\hat{\boldsymbol{u}}(\theta)$. Namely, the policy generalizing TP to curved spaces is based on the parallel transport equation [14]

$$\dot{\hat{u}}^k + \gamma^k_{ij} \dot{r}^i \hat{u}^j = 0\,, \tag{3.6}$$

which is to be integrated together with Eq. (3.1) and where the coefficients γ^k_{ij} are the Christoffel symbols associated with the metric tensor of the physical space $\boldsymbol{h}$.

The extension of ZP to motion on curved surfaces has instead already been worked out in Chap. 2. It essentially amounts to the solution (2.9), (2.13) and once again we will hereafter refer to it as the Randers Policy (RP).

3.3.1 Optimal Navigation on a Wavy Surface

As pictured in Fig. 3.7, we consider a one-dimensional wave-like perturbation of the 2D plane. Such a surface can be locally described, using the Monge representation introduced in Eq. (2.3), by the height function $z(x, y) = \Delta \sin(k_w x)$, where the parameters k_w and Δ set the period and amplitude of the wave (both in units of ℓ). From Eq. (2.4), the corresponding metric tensor then takes the form

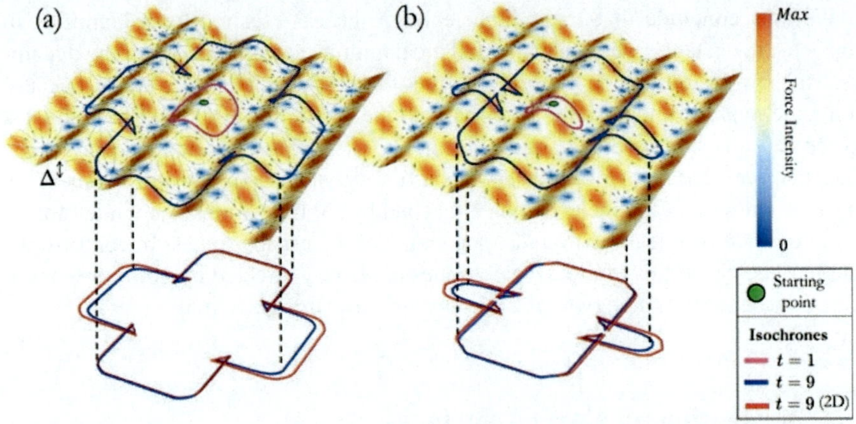

Fig. 3.7 Schematic representation of the wavy surface used to study the effect of space curvature on the active particle navigation in the same potential landscape defined by Eq. (3.2). Two short-time RP isochrones ($t = 1$ and $t = 9$ corresponding to the magenta and blue curves, respectively) are shown here for two different starting points (green dots): **a** a maximum and **b** a saddle point. The relative force strength is here fixed at $\gamma = 0.79$, while the wave amplitude is set to $\Delta = 0.5$. The most immediate effect of space curvature is to break the force field spatial symmetry. This is apparent by comparing the 2D projection of the isochrone on the surface with its counterpart obtained from the planar simulations ($\Delta = 0$, red curve). The former is indeed clearly shorter along the direction of the wave. In both panels the color gradient shows for the intensity of the external force field while the grey arrows indicate its direction

$$\boldsymbol{h}(x, y) = \begin{pmatrix} 1 + \Delta^2 k_w^2 \cos^2(k_w x) & 0 \\ 0 & 1 \end{pmatrix} . \tag{3.7}$$

As our numerical results do not show significant variations with k_w (not shown), we set $k_w = k$ so as to match the period of the wave with that of the potential (3.2).

A feature occurring in presence of nonzero space curvature is that the local force field strength $f = \sqrt{h_{ij} f^i f^j}$ depends on the properties of the metric $\boldsymbol{h}$, namely here $f^2 = {f^x}^2[1 + k_w^2 \Delta^2 \cos^2(k_w x)] + {f^y}^2$. In order to properly distinguish between the effects of the local curvature and that of the force amplitude, in the following we rescale the force field components in order to keep f independent of k_w and Δ. It is straightforward to show that for a general metric an appropriate rescaling corresponds to the following transformation to the force field components:

$$f^x \longmapsto \frac{1}{\sqrt{h_{xx}}} \left(f^x - f^y \frac{h_{xy}}{\sqrt{h}} \right), \qquad f^y \longmapsto f^y \sqrt{\frac{h_{xx}}{h}} ,$$

where h stands as a shorthand notation for $\det \boldsymbol{h}$. For the following analysis, we moreover focus on the most interesting confining regime by fixing $\gamma = 0.79$ while the performance of TP and RP will be assessed varying the wave amplitude Δ.

3.3.2 *Properties of the Isochrones in Presence of Space Curvature*

Figure 3.7 shows two short-time isochrones obtained from RP trajectories starting from a local maximum and saddle point of the potential. As shown from their projections on the plane, the shape of the isochrones is not strongly influenced by the presence of finite curvature. The curves are indeed not modified in the direction transverse to that of the surface wave, while they appear compressed along the wave direction. Here, the surface curvature therefore essentially introduces a breaking of the discrete rotational symmetry of the potential. Despite these quantitative differences with the planar case, the qualitative features of the RP isochrones such as the presence of self-intersections are unchanged by the presence of space curvature. As for the planar case, we did not find significant variations in the isochrones properties with the departure point, thus below we fix it to be a local maximum of the potential.

As shown in Fig. 3.8, the presence of surface oscillations does not qualitatively modify the isochrones properties for RP and TP. Indeed, the exponent α ruling the long-time growth of the isochrones area with time is in both cases constant upon varying Δ, and takes values $\alpha^{\mathrm{RP}} \approx 1$ and $\alpha^{\mathrm{TP}} \approx \frac{1}{2}$ (Fig. 3.8a). Moreover, the fractional area ϕ for RP oscillates around a well defined value ≈ 0.85 (Fig. 3.8b) while it decays to zero at long-times for TP (not shown). Analogously to the planar case and in contrast with TP, the optimized Randers policy allows the self-propelled particles to make use of the potential landscape in order to visit otherwise inaccessible regions, thus ensuring ergodic spatial exploration even in the presence of potential confinement. As the evaluation of geodesic distance is generally computationally demanding, we characterize the sensitivity of the policies via the scaling of the number of trajectories N_{out} lying on the outer isochrone. Consistently with the behaviour of the isochrone area, our results indicate that N_{out} decays exponentially with time for

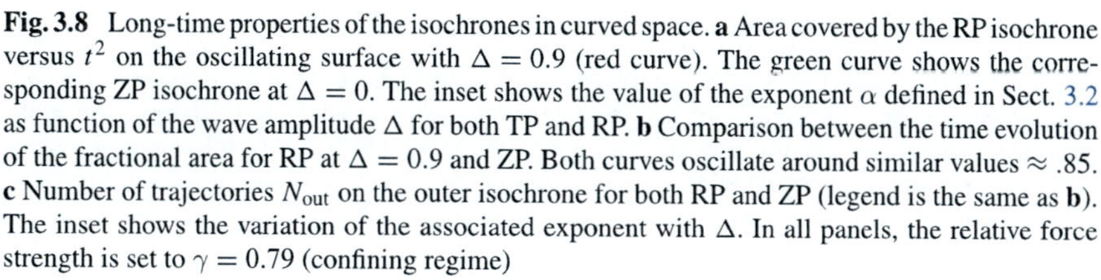

Fig. 3.8 Long-time properties of the isochrones in curved space. **a** Area covered by the RP isochrone versus t^2 on the oscillating surface with $\Delta = 0.9$ (red curve). The green curve shows the corresponding ZP isochrone at $\Delta = 0$. The inset shows the value of the exponent α defined in Sect. 3.2 as function of the wave amplitude Δ for both TP and RP. **b** Comparison between the time evolution of the fractional area for RP at $\Delta = 0.9$ and ZP. Both curves oscillate around similar values $\approx .85$. **c** Number of trajectories N_{out} on the outer isochrone for both RP and ZP (legend is the same as **b**). The inset shows the variation of the associated exponent with Δ. In all panels, the relative force strength is set to $\gamma = 0.79$ (confining regime)

RP (Fig. 3.8c) and remains constant for TP. The corresponding exponent λ_N moreover varies little with Δ, highlighting the robustness of the isochrones properties with space curvature.

3.4 Results Obtained in a Quasi-periodic Potential Landscape

In this chapter, we have studied an exemplary case for which navigation occurs in the presence of a sinusoidal potential landscape. Now, we shall provide evidence suggesting that the results obtained so far do not depend on the specific choice of the potential. In fact, the navigation strategy performances and robustness mostly rely on whether the external force field is confining or not. Therefore, in order to illustrate the generality of our results, let us consider the following quasi-periodic potential

$$U(\boldsymbol{r}) = -\frac{u}{4}\left[\cos(kx) + \cos(\sqrt{3}kx)/\sqrt{3} + \cos(ky) + \cos(\sqrt{3}ky)/\sqrt{3}\right], \quad (3.8)$$

which breaks translational symmetry as shown in the illustration of the corresponding force field in Fig. 3.9a. For the results presented below we work in the same units and with the same definitions γ as those defined in Sec. 3.1.

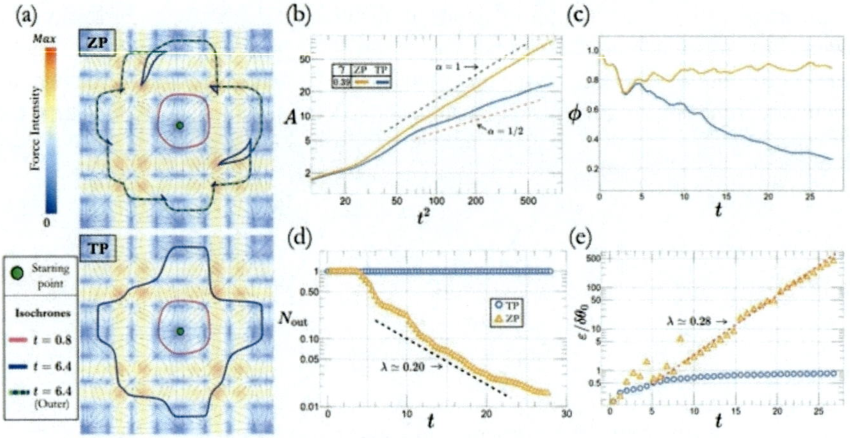

Fig. 3.9 Isochrones analysis in a 2D quasi-periodic potential landscape. **a** Each panel shows two exemplary short-time isochrones corresponding to $t = 0.8$ and $t = 6.4$. Here, the grey arrows indicate the external force field direction while the colour map shows its intensity. **b** Isochrones area as function of squared time for ZP and TP in the confining regime. **c** Fractional area covered by the isochrones as function of time, the legend is the same as **b**. **d** Number of trajectories located on the outer isochrone as function of time. The initial orientation resolution here is set to $\delta\theta_0 = 10^{-5}$. **e** Scaling with time of the mean angular separation ε for TP (blue circles) and ZP (orange triangles). Legend and parameters are the same as **d**. In all panels $\gamma = 0.39$

Figure 3.9 shows exemplary TP and ZP isochrones as well as the associated long-time scaling laws. In the regime where the force field overcomes the self-propulsion in finite regions, all the findings obtained with the periodic potential landscape still hold. Indeed, ZP proves to be still able to ergodically explore the space, as evidenced by the presence of self-crossings (Fig. 3.9a), as well as the scalings of the isochrone area $A \sim t^2$ (Fig. 3.9b) and the fractional area ϕ settling around a finite value (Fig. 3.9c), as opposed to TP. Lastly, both the number of trajectories N_{out} placed on the outermost part of the isochrones (Fig. 3.9d) and the mean angular separation (Fig. 3.9e) show an exponential trend for ZP. This confirms the strong sensitivity to initial conditions of this strategy and at the same time validates the robustness of TP for which no chaos is found.

3.5 Summary and Discussion

We have shown that the long-time properties of isochrones provide useful information about the efficiency of navigation strategies for exploration. Our results reveal that for a deterministic dynamics in presence of strong motion-limiting force the travel time minimizing strategies—ZP on the plane and RP in curved space—surprisingly allow for ergodic exploration, contrary to the Trivial Policy for which only a restricted portion of space is visited. Isochrones of ZP and RP are indeed nearly isotropic with a fractional area $\phi \geq 0.80$, while those of TP extend only along specific directions corresponding to escape routes drawn by force field landscape.

However, our results also indicate that ZP and RP lead to chaotic dynamics, which manifests as an exponentially fast divergence of trajectories whose number at the isochrones outer boundaries also decays exponentially. The resulting sensitivity to initial conditions is more pronounced in the confining regime, where it becomes increasingly harder to resolve the isochrones over long times. This complexity is however necessary to ensure full space coverage, as the only option for the particles to reach certain regions of space—where the external force is too strong—is to circumvent them.

More dramatically, ZP was found to be unstable to the presence of fluctuations in the dynamics, as those lead to a global trapping of trajectories regardless of the initial condition. If fluctuations cannot be neglected—as is often the case for the motion of microscopic objects—ZP, and by extension RP, are thus poor candidates for optimal navigation in presence of strong external forces.

These results moreover counter-intuitively suggest that, even though they are designed to maximize the effective speed of the particles, ZP and RP do not always constitute good escape strategies. Indeed, compared to ZP or RP, the trivial TP generally leads to a higher fraction of trajectories reaching a certain distance from the starting point, a difference that is even larger in presence of noise.

Although most of the results presented in this chapter were obtained with the sinusoidal potential (3.2), we expect our main conclusions to hold in more general settings. Indeed, so long as the potential is locally confining—the resulting external force overcomes self-propulsion only in disconnected regions—the ZP isochrones should always exhibit self-crossings. Moreover, in the long-time regime where their area is much larger than the scale associated with the potential, it is natural to expect that the behavior of the isochrones will remain qualitatively independent of the details of the potential. To stress this idea, we have indeed provided in Sect. 3.4 simulation results obtained with a quasi-periodic potential in the confining regime which show that all scaling laws and measures of chaos reported for the sinusoidal potential remain qualitatively unchanged.

In light of the results reported here, notions of stability could be a decisive factor for the design of navigation strategies for ballistic exploration of rugged landscapes in order to ensure reasonable performances. More generally, our results highlight an emergent trade-off between exploration efficiency and stability, which, depending on the context and desired task, must be taken into consideration for the design of the corresponding navigation policy.

References

1. Piro L, Golestanian R, Mahault B (2022) Efficiency of navigation strategies for active particles in rugged landscapes. Front Phys 10. ISSN 2296-424X. https://www.frontiersin.org/articles/10.3389/fphy.2022.1034267
2. Liebchen B, Löwen H (2019) Optimal navigation strategies for active particles. EPL (Europhys Lett) 127(3):34003. https://doi.org/10.1209/0295-5075/127/34003
3. Daddi-Moussa-Ider A, Löwen H, Liebchen B (2021) Hydrodynamics can determine the optimal route for microswimmer navigation. Commun Phys 4(1):15. https://doi.org/10.1038/s42005-021-00522-6
4. Piro L, Tang E, Golestanian R (2021) Optimal navigation strategies for microswimmers on curved manifolds. Phys Rev Res 3:023125. https://link.aps.org/doi/10.1103/PhysRevResearch.3.023125
5. Bao D, Robles C, Shen Z (004) Zermelo navigation on Riemannian manifolds. J Differ Geom 66(3):377 – 435. https://doi.org/10.4310/jdg/1098137838
6. Piro L, Mahault B, Golestanian R (2022) Optimal navigation of microswimmers in complex and noisy environments. New J Phys 24(9):093037. https://doi.org/10.1088/1367-2630/ac9079
7. Biferale L, Bonaccorso F, Buzzicotti M, Clark Di Leoni P, Gustavsson K (2019) Zermelo's problem: optimal point-to-point navigation in 2d turbulent flows using reinforcement learning. Chaos 29:103138. https://doi.org/10.1063/1.5120370
8. Wolfram Research Inc. (2022) Mathematica, Version 13.1.0. https://www.wolfram.com/mathematica
9. Zarfaty L, Peletskyi A, Fouxon I, Denisov S, Barkai E (2018) Dispersion of particles in an infinite-horizon lorentz gas. Phys Rev E 98:010101. https://link.aps.org/doi/10.1103/PhysRevE.98.010101
10. Chitsaz H, LaValle SM (2007) Time-optimal paths for a dubins airplane. In: Proceedings of the 46th IEEE conference on decision and control 2007, CDC, Proceedings of the ieee conference on decision and control. Institute of Electrical and Electronics Engineers Inc., United States, pp 2379–2384. ISBN 1424414989. https://doi.org/10.1109/CDC.2007.4434966. 46th IEEE Conference on Decision and Control 2007, CDC ; Conference date: 12-12-2007 Through 14-12-2007
11. Guerrero JA, Bestaoui Y (2013) UAV path planning for structure inspection in windy environments. J Intell Robot Syst 69:297–311. https://doi.org/10.1007/s10846-012-9778-2

12. Lin S-Z, Li Y, Ji J, Li B, Feng X-Q (2020) Collective dynamics of coherent motile cells on curved surfaces. Soft Matter 16:2941–2952. http://dx.doi.org/10.1039/C9SM02375E
13. Callens SJP, Uyttendaele RJC, Fratila-Apachitei LE, Zadpoor AA (2020) Substrate curvature as a cue to guide spatiotemporal cell and tissue organization. Biomaterials 232:119739. ISSN 0142-9612. https://doi.org/10.1016/j.biomaterials.2019.119739. http://www.sciencedirect.com/science/article/pii/S0142961219308579
14. Deserno M (2014) Notes on differential geometry. https://www.cmu.edu/biolphys/deserno/pdf/diff_geom.pdf

Chapter 4
Optimal Navigation in Complex and Noisy Environments

The real point of honor [for a scientist] is not to be always right. It is to dare to propose new ideas, and then to check them.

Pierre-Gilles de Gennes

Part of this chapter (Sects. 4.1–4.3) has been adapted from Piro L., Mahault B., Golestanian R., *New J. Phys.*, **24** 093037, (2022) [1]. I have taken part in the conception of the research and in the redaction of the manuscript. I have performed the numerical simulations and analyzed the corresponding data. I have also participated in the development of the simulation code.

As we have already discussed in Sects. 1.2.1 and 1.3.1, optimal navigation in the presence of noise falls into the class of problems addressed by stochastic optimal control (SOC) theory [2–4], while at the same time, reinforcement learning (RL) algorithms can provide convenient and increasingly popular routes to determining the optimal control landscapes [5–7]. In practice, implementing these policies can be achieved via external feedback loops such as shown in Fig. 4.1c for the actuation of a microswimmer motion [8–11].

On the other hand, a number of natural and artificial microswimmers exhibit tactic behaviour [12], i.e. are able to adapt their motility in response to external stimuli such as light [13, 14], chemical concentration [15, 16] or viscosity [17] gradients in an autonomous fashion (Fig. 4.1a). Harnessing guidance provided by taxis allows microswimmers to perform complex tasks [18–20] in a semi-autonomous way as it can be carried out without the real-time intervention of an external operator (Fig. 4.1b).

Here, we show how these ideas can be applied to the problem of optimal navigation in complex and noisy environments. Considering a minimal but non-trivial optimization problem in two dimensions, we note that the stochastic trajectories obtained from the optimal control maps are closely distributed around the path defined by Zermelo's approach, which is optimal at zero-noise. Building on this observation, we propose novel navigation policies that consist for the active particle in regulating

L. Piro, *Optimal Navigation in Active Matter*, Springer Theses,
https://doi.org/10.1007/978-3-031-52577-3_4

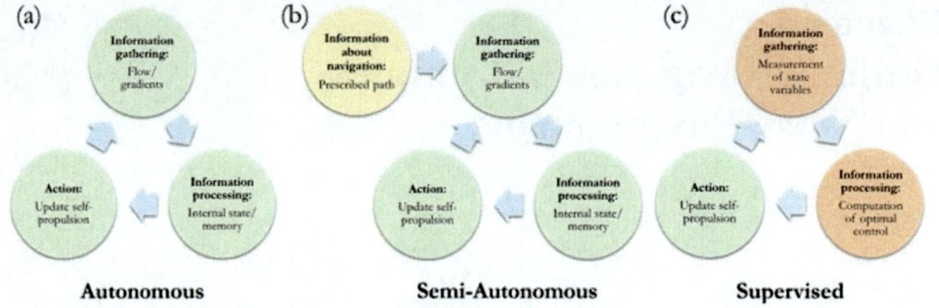

Fig. 4.1 Autonomous versus supervised optimal navigation. **a** An active particle can navigate fully autonomously by employing its sensing apparatus to collect information about its local environment, process it internally, and update its motility machinery accordingly. As illustrated in **b**, the implementation of semi-autonomous navigation strategies is based on a similar feedback cycle, except that now the information about the optimal path is provided by an external operator. **c** Supervised navigation, on the contrary, requires the intervention of an external interpreter (here represented by the red color) within the feedback cycle in order to gather the relevant information on the swimmer state and use it to determine the optimal behavior

its relative position to Zermelo's path. As the latter can be detected by an active agent if highlighted by a stationary field (e.g. by applying a light (resp. chemical) gradient for a phototactic (resp. chemotactic) particle), the protocols we propose can be implemented in a semi-autonomous fashion. Moreover, confining the particle around Zermelo's path improves the stability of its motion, which can be crucial for the particle navigation efficiency and robustness, as pointed out in Chap. 3.

Using extensive Brownian dynamics simulations, we show that the new policies show arrival time statistics comparable to that obtained from stochastic optimal control, resulting in excellent performances. The new policies are moreover found to be robust upon changes in the environment as well as positional and rotational fluctuations, while being easily generalizable to more complex navigation problems such as navigation on curved manifolds or in the presence of random flows. Lastly, we introduce a toy model that allows for a more systematic study of the protocols, thus providing valuable insights into their performance in different dynamical regimes.

4.1 From Supervised to Semi-autonomous Optimal Navigation

We consider an overdamped self-propelled particle moving at a fixed speed v_0 in presence of a stationary force field $\boldsymbol{f}(\boldsymbol{r})$, which may in general include a contribution due to advection by the solvent flow velocity, and translational diffusion with diffusivity D. For simplicity, we set the friction coefficient (and mobility) to unity. The position $\boldsymbol{r}$ of the self-propelled particle thus obeys Eq. (1.26), which we shall conveniently report here:

$$\dot{\boldsymbol{r}} = v_0 \hat{\boldsymbol{u}} + \boldsymbol{f}(\boldsymbol{r}) + \sqrt{2D}\,\boldsymbol{\xi}\,, \tag{4.1}$$

where $\hat{\boldsymbol{u}}$ is the unit vector setting the direction of self-propulsion, and $\boldsymbol{\xi}$ is a Gaussian white noise vector whose components have unit variance. Within this setting, the only degree of freedom accessible to the self-propelled particle for navigation is its orientation $\hat{\boldsymbol{u}}$. For the sake of presentation, we will first assume a full control over $\hat{\boldsymbol{u}}$, either from external sources or by the particle itself, while this constraint will be relaxed later.

We now want to determine the optimal navigation protocol for the particle to reach a target position $\boldsymbol{r}_T$ in the shortest time (on average) given a stationary force profile and an initial position $\boldsymbol{r}$. To this end, we can refer to Sect. 1.3 where we have shown that this can be obtained from the stochastic HJB equation for the mean first-passage time $\mathcal{T}(\boldsymbol{r})$ (MFPT), namely

$$\begin{cases} -v_0|\nabla\mathcal{T}| + \boldsymbol{f}(\boldsymbol{r})\cdot\nabla\mathcal{T} + D\nabla^2\mathcal{T} = -1 \\ \mathcal{T}(\boldsymbol{r}_\mathrm{T}) = 0\,. \end{cases} \tag{4.2}$$

The solution of this partial differential equation indeed allows to determine the optimal control map via

$$\hat{\boldsymbol{u}}_\mathrm{opt}(\theta(\boldsymbol{r})) = -\frac{\nabla\mathcal{T}}{|\nabla\mathcal{T}|}\,, \tag{4.3}$$

which defines what we call the Optimal Policy (OP).

Although in the limit of vanishing diffusivity OP is equivalent to Zermelo's Policy (ZP), their implementation requires different amount of knowledge of the environment. Indeed, Zermelo's path (optimal at $D = 0$) can be integrated in an autonomous way assuming the particle able to measure the local force and its gradients (Fig. 4.1a), whereas OP formally requires full knowledge of the global control map $\hat{\boldsymbol{u}}_\mathrm{opt}(\theta(\boldsymbol{r}))$ obtained from Eqs. (4.2–4.3) which is generally achieved in a supervised manner (Fig. 4.1c).

We now illustrate OP by considering a simple but nontrivial setup in which a self-propelled particle navigates in the two dimensional plane spanned by the unit vectors $\{\hat{\boldsymbol{e}}_x, \hat{\boldsymbol{e}}_y\}$ between neighbouring stationary points of a Taylor-Green vortex flow (see the colour map and black arrows in Fig. 4.2a). This configuration corresponds to $\boldsymbol{f}(\boldsymbol{r}) = v_\mathrm{f}[\cos(ky)\sin(kx)\,\hat{\boldsymbol{e}}_x - \cos(kx)\sin(ky)\,\hat{\boldsymbol{e}}_y]$ with $k = 2\pi/\ell$ and ℓ being the characteristic length scale of the flow. Rescaling space and time as $\boldsymbol{r} \to \ell\boldsymbol{r}$ and $t \to \ell t/v_0$, the dynamics (4.1) is characterized by only two nondimensional parameters: the ratio between flow intensity and self propulsion $\gamma \equiv v_\mathrm{f}/v_0$, and the Péclet number[1] $\mathrm{Pe} = \ell v_0/D$. Here, we will focus only on cases where the self-propulsion is always stronger than the flow, namely, $0 \le \gamma \le 1$.

---

[1] This definition of Péclet number differs from the one of an ABP in Eq. (1.5) since we have here used the flow length scale (in place of the microswimmer size) to non-dimensionalize the system dynamics.

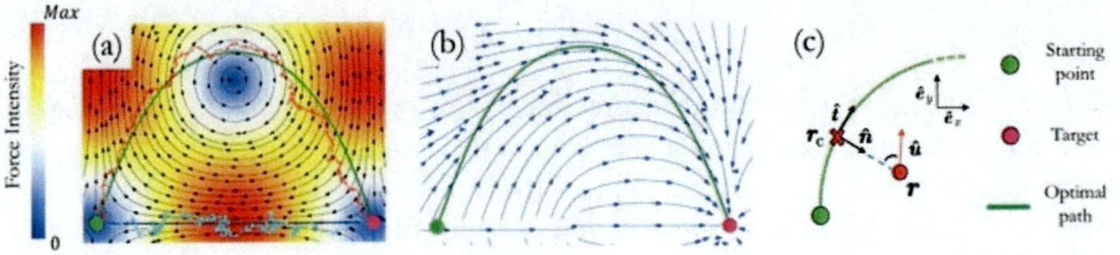

Fig. 4.2 Optimal navigation in a Taylor-Green flow. **a** The Taylor-Green flow map, where the colour codes for the intensity and the black arrows indicate the direction. The green curve indicates Zermelo's optimal trajectory at vanishing noise connecting the initial point $\boldsymbol{r}_0 = -(\ell/2)\hat{\boldsymbol{e}}_x$ and the target $\boldsymbol{r}_\mathrm{T} = \boldsymbol{0}$, marked with a green and magenta circle, respectively. The red and cyan lines show two representative stochastic trajectories following respectively the Optimal and Straight policies at $\gamma = 0.7$ and $\mathrm{Pe} = 400$. **b** The optimal control obtained by solving Eqs. (4.2) and (4.3) for same parameters as in panel (**a**). **c** Schematic defining the quantities used for the implementation of the semi-autonomous navigation strategies. The red dot indicates the active particle position $\boldsymbol{r}$ and the red arrow its heading direction $\hat{\boldsymbol{u}}$. $\boldsymbol{r}_c$ (red cross) marks the closest point to $\boldsymbol{r}$ on the optimal trajectory, while the two vectors $\hat{\boldsymbol{n}}$ and $\hat{\boldsymbol{t}}$ are respectively normal and tangent to the optimal trajectory in $\boldsymbol{r}_c$

With the setup shown in Fig. 4.2a, the most direct route between the departure and arrival points requires travelling counter-flow all the way. Consequently, the straight path becomes increasingly disadvantageous as the flow amplitude grows, such that for $\gamma \gtrsim 0.4$ Zermelo's path makes use of the flow profile and takes a bell-shaped curve (see the green line in Fig. 4.2a). For finite Péclet number ($D \neq 0$), such a feature is moreover consistent with the control map provided by OP, since the latter generally orients the self-propulsion away from the straight path (Fig. 4.2b). Simulations of the Brownian dynamics (4.1) with the control map solving Eqs. (4.3) and (4.2) indeed reveal that the OP trajectories tend to remain close to Zermelo's path for a broad range of Péclet number values, provided that the flow strength and particle self-propulsion dominate over fluctuations (see e.g. the density map in Fig. 4.3d and Appendix A.1 for more details on the stochastic dynamics simulations).

The above observations suggest that optimized navigation in the finite Péclet number regime may be achievable using only the local information of the relative positions of the stochastic swimmer and Zermelo's path, as opposed to the OP which requires the knowledge of a global control map $\sim \nabla \mathcal{T}(\boldsymbol{r})$.

We now explore a number of such local strategies and probe their efficiencies in comparison with OP. For a given particle position $\boldsymbol{r}$, we define $\boldsymbol{r}_c \equiv \min_{\boldsymbol{r}'} |\boldsymbol{r} - \boldsymbol{r}'|$ as the corresponding closest point on Zermelo's path. Moreover, we suppose that the latter is smooth and can be parametrized by the moving frame $\{\hat{\boldsymbol{t}}, \hat{\boldsymbol{n}}\}$, of tangent and normal vectors, with $\hat{\boldsymbol{t}}$ heading towards the target as shown in Fig. 4.2c. Assuming that the swimmer is able to measure its relative position to Zermelo's path, it can regulate it by steering its self propulsion direction $\hat{\boldsymbol{u}}$ via the following rule

$$\hat{\boldsymbol{u}} \cdot \hat{\boldsymbol{n}} = \mathcal{G}(\Delta r, \boldsymbol{f}(\boldsymbol{r})) \,, \tag{4.4}$$

where $\Delta r \equiv (\boldsymbol{r} - \boldsymbol{r}_c) \cdot \hat{\boldsymbol{n}}$ and the function $\mathcal{G} \in [-1, 1]$ depends on the amount of information available to the swimmer. As the rhs of Eq. (4.4) depends on $\boldsymbol{r}$ solely

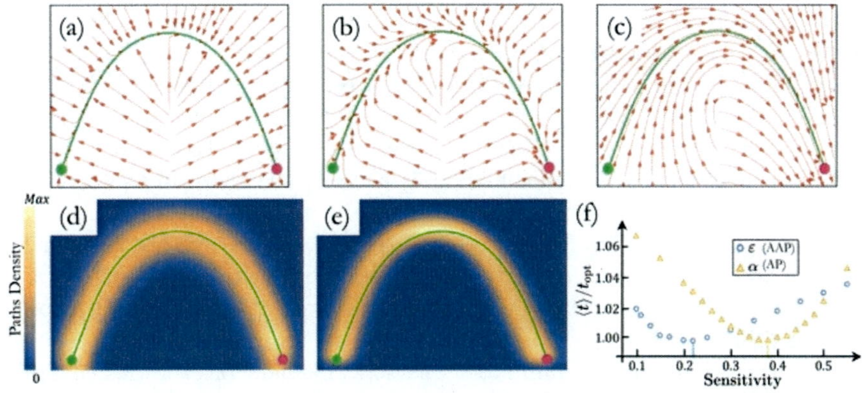

Fig. 4.3 The semi-autonomous navigation policies. **a–c** The control $\hat{\boldsymbol{u}}$ obtained for the AAP from Eqs. (4.4) and (4.6) in the Taylor-Green flow at $\gamma = 0.7$ with $\varepsilon = 0.05$ **(a)**, 0.2 **(b)** and 1 **(c)**. **d, e** Heat maps of 10^3 stochastic trajectories obtained from numerical simulations of the OP **(d)** and AAP **(e)**. The solid green lines in **a–e** represent the $D = 0$ Zermelo's path. **f** Example of curves leading to the determination of the optimal sensitivities for the AP and AAP ($\gamma = 0.7$ and $\mathrm{Pe} = 400$) with $\langle t \rangle$ and t_{opt} respectively denoting the mean arrival time and the optimal travel time in absence of noise. The optimal values of α(AP) and ε(AAP) are indicated by the vertical dashed lines. Note the small range of values for the normalized mean arrival time, showing the robustness of the two policies. Here, both curves are normalized by their minimum value

through Δr and the external flow $\boldsymbol{f}$, it defines a class of optimal navigation policies whose implementation relies only on the swimmer's local knowledge of its environment.

In the simplest case where the swimmer can only determine the direction $\hat{\boldsymbol{n}}$ to Zermelo's path (from its current position), it can choose to keep a constant angle α between its self-propulsion direction and $\hat{\boldsymbol{n}}$. Such *Aligning Policy* (AP) corresponds to a protocol

$$\mathcal{G}_{\mathrm{AP}} = \pm\cos(\alpha)\,, \tag{4.5}$$

where the $\pm$ sign ensures that $\hat{\boldsymbol{u}} \cdot \hat{\boldsymbol{t}} \geq 0$. We moreover note that Eq. (4.5) defines a strategy resembling the constant bearing pursuit [21, 22] of a target located at position $\boldsymbol{r}_c$.

Although AP allows the swimmer to remain in the vicinity of Zermelo's path, it also slows it down as it imposes a finite angle between $\hat{\boldsymbol{u}}$ and $\hat{\boldsymbol{t}}$ even for (arbitrarily) small separations. For swimmers able to evaluate their relative distances to Zermelo's path, AP can thus be refined by allowing $\mathcal{G}$ to depend on Δr. This defines the *Adaptive Aligning Policy* (AAP). Here, we choose, for simplicity, a piecewise linear form for $\mathcal{G}(\Delta r)$, namely

$$\mathcal{G}_{\mathrm{AAP}}(\Delta r) = \begin{cases} +1 & \text{if} \quad \Delta r < -\varepsilon \\ -\Delta r/\varepsilon & \text{if} \quad |\Delta r| < \varepsilon \\ -1 & \text{if} \quad \Delta r > \varepsilon \end{cases}, \tag{4.6}$$

where the parameter ε sets a cut-off scale above which the stochastic particle points normally to Zermelo's path.

The parameters α and ε introduced above essentially play the role of sensitivities for AP and AAP, respectively. Their optimal values (that minimize the mean travel time in this case) cannot be selected *a priori*, and need to be determined empirically. However, the existence of such optimal values can be intuitively understood from the control maps obtained from Eq. (4.6) for various ε values (with the generalization to AP being straightforward). As shown in Fig. 4.3a–c, exceedingly small ε values force the swimmer to mostly point normally to Zermelo's path, while for excessively large ε stochastic trajectories are less efficiently confined and can visit less favourable flow regions. Therefore, it is natural to expect an intermediate value of ε providing the optimal trade-off between efficient confinement and tangential motion along Zermelo's path.

This heuristic picture is confirmed by numerical simulations showing that the mean arrival time $\langle t \rangle$ indeed exhibits a minimum at a value $\varepsilon = \varepsilon_{\text{opt}}$ (see Fig. 4.3f). We moreover note that $\langle t \rangle$ varies relatively little with ε, such that in practice the policy implementation does not require a fine tuning of this parameter. The heat map of trajectories obtained from simulations of AAP at optimal ε shows that they globally follow Zermelo's path (Fig. 4.3e), similarly to the OP case.

Contrary to OP, however, the stochastic trajectories are not distributed symmetrically with respect to the desired path due to a non-zero transverse component of the force (see Fig. 4.2a). Closer agreement with the OP results and improved protocol stability thus requires additional features such as a policy function $\mathcal{G}$ in Eq. (4.4) that depends explicitly on the local force field $\boldsymbol{f}(\boldsymbol{r})$. In particular, compensating for the transverse drift from the force could lead to protocols similar to well-established control strategies such as Sliding Mode Control [23]. Nevertheless, we focus for the moment on the case where the swimmer is unaware of the local flow structure around it and will address such more elaborate policies in Sect. 4.4.

4.2 Performance Assessment of the Navigation Policies

4.2.1 Arrival Time Statistics

We now compare the performances of the two policies introduced above (AP and AAP) with that of OP by simulating Eq. (4.1) with the controls defined by (4.3) and (4.4) in the Taylor-Green flow setup (Fig. 4.2a). To illustrate the relevance of nontrivial policies, we moreover consider the Straight Policy (SP)—introduced in Sect. 2.4– for which the swimmer always points towards the direction of the target regardless of its current position (see the cyan curve in Fig. 4.2a for a representative trajectory).

We first work at fixed $\text{Pe} = 400$ and vary the relative flow amplitude $\gamma \in [0, 1]$. Figure 4.4a shows the arrival time distributions $P(\tau)$ with $\tau \equiv t/\langle t \rangle$ for each of the four policies (OP, AP, AAP, and SP) at $\gamma = 0.7$. Remarkable overlap between the OP and AAP distributions can be observed. We moreover find that they are both well described by a so-called inverse Gaussian distribution of the form

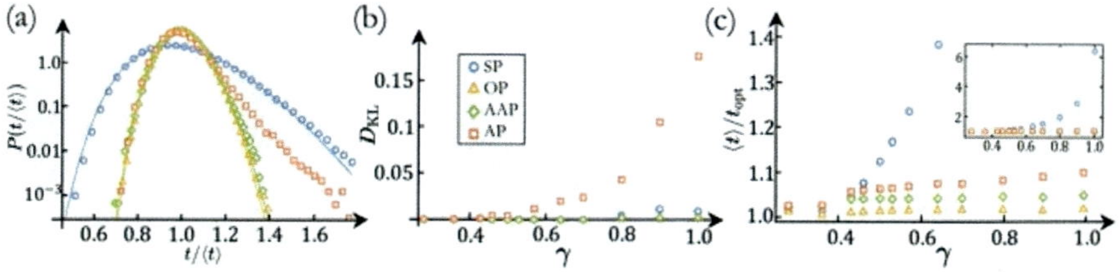

Fig. 4.4 Comparison of the performances of the different policies. **a** Arrival time distribution as a function of the normalized time $t/\langle t\rangle$ for the four navigation strategies: Straight Policy (blue circles), Optimal Policy (yellow triangles), Adaptive Aligning Policy (green diamonds) and Aligning Policy (red squares). **b** Kullback-Leibler divergence D_{KL} between the numerical and theoretical (Eq. (4.7)) distributions as a function of the relative flow amplitude γ. **c** Mean arrival time (normalized by the noiseless optimal time t_{opt}) as a function of γ. Inset: Zoom-out showing the larger values taken by SP. In **a–c** the symbols show the data obtained from Langevin simulations, while in **a** the solid lines indicate the theoretical prediction (4.7) calculated with $\langle t\rangle$ and σ obtained from the numerical data. All data in **a–c** are averaged over 10^5 independent trajectories

$$P(\tau) = \sqrt{\frac{\langle t\rangle^2}{2\pi\sigma^2\tau^3}}\exp\left[-\frac{\langle t\rangle^2(\tau-1)^2}{2\sigma^2\tau}\right], \tag{4.7}$$

with variance σ^2. To quantify this correspondence we furthermore calculate the Kullback-Leibler divergence $D_{\mathrm{KL}} = \langle \ln[P_{\mathrm{num}}(\tau)/P(\tau)]\rangle_{P_{\mathrm{num}}}$ between the numerically obtained distribution P_{num} and the prediction of Eq. (4.7), with $\langle t\rangle$ and σ determined from the data. As shown in Fig. 4.4b, D_{KL} remains almost zero for both OP and AAP over a wide range of γ values, highlighting the robustness of (4.7). As the inverse Gaussian corresponds to the First Passage Time distribution of a driven Brownian particle in one dimension [24], Eq. (4.7) is closely related to the confinement of the OP and AAP trajectories along Zermelo's path as shown in Fig. 4.3d, e. In fact, for both OP and AAP the loss of correspondence with the inverse Gaussian coincides with the regime of strong fluctuations that prevent the swimmers from being efficiently guided along Zermelo's path (details in Sect. 4.2.2). Interestingly, SP also shows arrival time distributions globally compatible with the inverse Gaussian (4.7), with only slight deviations at large flow amplitudes (see blue circles in Fig. 4.4a, b), indicating that in this case too trajectories are nearly one dimensional. As SP trajectories are mostly oriented against the flow, they are characterized by a lower effective drift on average, resulting in a larger ratio $\sigma/\langle t\rangle$.

In contrast, for sufficiently large flow amplitudes the simpler Aligning Policy shows arrival time distributions that do not follow the inverse Gaussian law (red squares in Fig. 4.4a, b). These distributions indeed exhibit a crossover from inverse Gaussian-like behaviour at $\tau < 1$ to an exponential decay at $\tau > 1$ with a characteristic time τ_{AP} that is systematically larger than the value $2\sigma^2/\langle t\rangle^2$ predicted by Eq. (4.7) (see Fig. 4.5b). Defining $\tau^* \equiv 2\sigma^2/\langle t\rangle^2$ as the variance to square mean ratio of arrival time for each protocol, its scaling with the flow strength is shown in Fig. 4.5a for all policies. As a sign that AAP and OP on average lead the swimmer

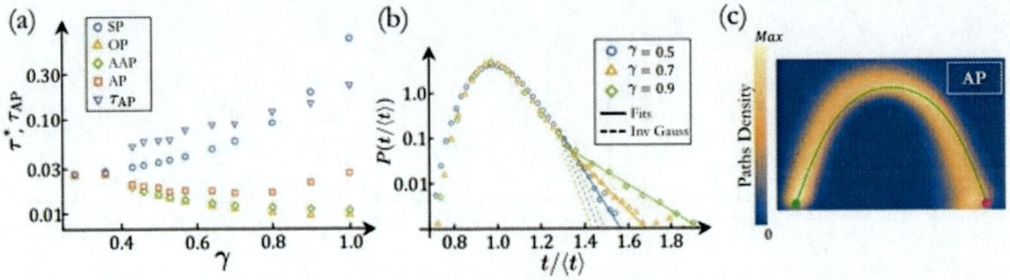

Fig. 4.5 **a** Arrival time variance to square mean ratio as function of the relative flow strength γ at Pe $= 400$ for the four policies: Straight Policy (blue circles), Optimal Policy (yellow triangles), Adaptive Aligning Policy (green diamonds) and Aligning Policy (red squares). The inverted triangles are the values obtained from the fit of the large-time tails of AP distribution (vertical axis in log scale). **b** Arrival time probability distributions as function of the normalized time $t/\langle t\rangle$ corresponding to AP for three different values of the relative flow strength. Solid lines show the exponential fit at large times, while the dashed lines correspond to the tails predicted by the inverse Gaussian law (4.7) using the values $\langle t\rangle$ and σ from the data. **c** Heat map of 10^3 stochastic trajectories obtained from numerical simulations at $\gamma = 0.7$ and Pe $= 400$ of AP. The green curve shows the Zermelo path connecting the initial point $\boldsymbol{r}_0 = -\ell/2\hat{\boldsymbol{e}}_x$ (green circle) and the target $\boldsymbol{r}_\mathrm{T} = \boldsymbol{0}$ (magenta circle). All data in **a** and **b** are averaged over 10^5 independent trajectories

to travel faster as the flow strength is increased, τ^* decays with γ for these two policies. Conversely, swimmers following SP always travel counter-flow and are thus get slower on average as γ increases. They are thus more subject to fluctuations, such that for SP τ^* grows with γ. Lastly, for AP τ^* undergoes a crossover from a decay with flow strength at small γ, to a growth with γ at large flows (red squares in Fig. 4.5a).

On the other hand, the value τ_AP obtained from the large-time tails of the distribution always grows with γ, similarly to SP (purple inverted triangles in Fig. 4.5a). These observations can be rationalized from the heat map of AP trajectories shown in Fig. 4.5c. Namely, it shows that most of the trajectories end at the left of the target, such that the swimmers have to travel counter-flow and are thus generally slower in the final stretch. This suggests that, for AP, the events characterized by $\tau > 1$ are dominated by this last part of the swimmer's journey, ultimately caused by the overall pronounced asymmetry with respect to the Zermelo path.

As customary in this context, we compare the performances of the navigation policies by measuring the mean arrival time to reach the target $\langle t\rangle$, here normalized using the optimal value t_opt at $D = 0$ obtained from Zermelo's solution. The corresponding results shown in Fig. 4.4c reveal that, naturally, OP performs the best with the mean arrival times always remaining higher than t_opt by only a few percent. Conversely, the trivial SP performances strongly deteriorate as γ increases. For small flow amplitudes where Zermelo's path is almost straight ($\gamma \lesssim 0.4$) SP performs similarly to OP with $\langle t\rangle \gtrsim t_\mathrm{opt}$, whereas for sufficiently large γ values it exhibits mean arrival times reaching five to six times t_opt (see the inset of Fig. 4.4c). Despite the presence of fluctuations, the performances of the policies are thus primarily set by their ability to make efficient use of the stationary flow profile. This feature is moreover illustrated by both AP and AAP, which show mean arrival times no more than 10% higher than that of OP, regardless of the relative flow amplitude. As expected, in the non-trivial cases ($\gamma > 0.4$) the performances of the different strategies reflect the amount of information they require for navigation, such that, in the order of increasing efficiency, one finds SP, AP, AAP and OP.

4.2.2 Robustness of the New Protocols

The above analysis shows that AAP displays arrival time statistics similar to that of OP. Both AP and AAP moreover exhibit performances comparable to OP, despite them relying only on local information. We now assess the generality of these results, focusing on AAP, by discussing more general situations with different model parameters and evaluation setups.

Increasing the Noise Strength

The two colour maps of Fig. 4.6a show how the ratio $\langle t \rangle / t_{\rm opt}$ varies with the Péclet number and relative flow amplitude γ for OP and AAP. In agreement with previous results, $\langle t \rangle / t_{\rm opt}$ for OP does not significantly depend on γ while we observe a slight increase with decreasing Pe. The AAP case, on the other hand, exhibits two distinct regimes. At small flow strengths ($\gamma \lesssim 0.4$), $\langle t \rangle / t_{\rm opt}$ remains nearly constant upon varying Pe such that the AAP performances are not significantly altered by the amplitude of noise. Conversely, at larger γ values where Zermelo's path is more curved, the mean arrival time is more affected by translational noise. As shown in Fig. 4.6b for $\gamma = 0.7$, all non-trivial strategies show a slight decrease in performance as Pe is lowered, whereas SP becomes slightly more favourable upon increasing the noise, since in this case stronger fluctuations lead the swimmer to visit less unfavourable flow regions.

Misalignment of Self-propulsion

We have so far assumed full control over the self-propulsion orientation $\hat{\boldsymbol{u}}$. In reality, however, $\hat{\boldsymbol{u}}$ is subject to fluctuations—e.g. due to rotational Brownian motion or inaccuracies in the evaluation of the desired direction—which affect the performances of the policies using it as a control. To model the effect of rotational noise, we applied random rotations $\hat{\boldsymbol{u}} \to \mathcal{R}(\beta)\hat{\boldsymbol{u}}$ to the controls (Eqs. (4.3) and (4.4)), where the angle β was sampled from a uniform distribution in $(-\eta\pi, \eta\pi]$; we present the corresponding results in Fig. 4.6c. For small η values, we find that the inverse of the mean arrival time normalized by its value at $\eta = 0$, $\langle t \rangle / \langle t \rangle_\eta$, decays linearly with η^2 with a policy-dependent slope. In particular, OP and AAP show similar trends and appear to be much more robust to the effect of rotational noise than the SP case, as for the latter the mean arrival time has increased by a factor 10 at $\eta = \frac{1}{2}$, while the corresponding drop in performance for OP and AAP is about 40%.

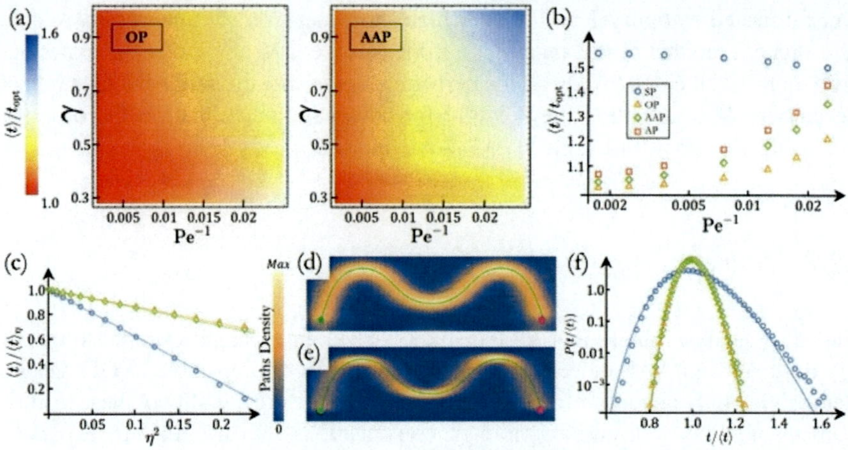

Fig. 4.6 Robustness of the navigation policies. **a** Colour maps showing the performances of OP (left) and AAP (right) as functions of the relative flow amplitude γ and the Péclet number Pe. **b** Normalized mean arrival time as a function of the Péclet number at $\gamma = 0.7$ for the four policies (horizontal axis in log scale). **c** Inverse of the mean arrival time as a function of the rotational noise strength η^2 at $\text{Pe} = 400$ and $\gamma = 0.7$ for OP, AAP and SP (caption is the same as in **b**). Here the data is normalized by the value $\langle t\rangle$ measured at $\eta = 0$. Solid lines show linear fits in the small η regime. **(d)**, **(e)** Heat maps of 10^3 stochastic trajectories obtained from numerical simulations at $\gamma = 0.7$ and $\text{Pe} = 400$ of OP and AAP, respectively. In both panels the green curve shows Zermelo's path connecting the initial point $\boldsymbol{r}_0 = -(3\ell/2)\hat{\boldsymbol{e}}_x$ (green circle) and the target $\boldsymbol{r}_\text{T} = \boldsymbol{0}$ (magenta circle). **f** Arrival time probability distributions as functions of $t/\langle t\rangle$ corresponding to (d,e) as well as for the SP case under similar conditions; the caption is the same as **b**. Solid lines show the theoretical curves obtained similarly to Fig. 4.4a. All data in **a–c** and (f) are averaged over 10^5 independent trajectories

The scaling $\langle t\rangle/\langle t\rangle_\eta - 1 \sim \eta^2$ can be understood from an effective one dimensional model of driven Brownian motion. Assuming that the particle remains in the vicinity of the mean path and neglecting effects due to the curvature of the latter, we consider the following one-dimensional dynamics

$$\dot{r}_\parallel = v_\parallel(r_\parallel)\cos\theta + \sqrt{2D}\xi_\parallel, \tag{4.8}$$

where the subscript $\parallel$ stands for quantities projected along the mean path. The first term on the rhs of (4.8) thus accounts for the projection of the total velocity, which includes the combined effects of flow and self-propulsion. In general, the angle θ obeys a nontrivial and policy-dependent dynamics. However, in the limit of small η and D where the particle remains close to the mean path, we approximate θ as a Gaussian noise with zero mean and variance $\propto \eta^2$. Expanding the cosine and performing an average over the noises therefore leads to $\langle \dot{r}_\parallel\rangle \simeq v_\parallel(r_\parallel)(1 - \kappa\eta^2)$ with $\kappa > 0$ being a constant that depends on the navigation details, e.g. the protocol used. In the one-dimensional approximation and assuming that $v_\parallel$ varies little with $r_\parallel$, the mean travelling time to reach an absorbing barrier at distance L scales as

$\langle t \rangle_\eta \sim L/\langle \dot{r}_\parallel \rangle$, such that we recover $\langle t \rangle / \langle t \rangle_\eta \simeq 1 - \kappa \eta^2$, which corresponds to the scaling observed in Fig. 4.6c. This scaling is thus not expected to hold for large noises, as suggested by the deviations from the linear decay observed for the largest η values in Fig. 4.6c.

Complex Navigation Tasks

Increasing the distance between the initial and target points in the Taylor-Green flow allows us to design more complex paths. Upon translating the initial swimmer position along $\hat{\boldsymbol{e}}_x$, as shown in Fig. 4.6d and e, both OP and AAP lead to trajectories focused around Zermelo's path on average. Consequently, the corresponding arrival time distributions remain well characterized by the inverse Gaussian law (Fig. 4.6f). The performances of AAP and OP are moreover stable upon increasing the total travel distance, with mean arrival times $\langle t \rangle$ not higher than t_{opt} by more than a few percent.

Large Fluctuations and the Inverse Gaussian Law

The correspondence observed between the inverse Gaussian law and the arrival time distributions of OP, AAP and SP is due to the fact that, as long as the thermal fluctuations are reasonably small, these policies guide the particle along a fictional path, thus reducing the system to an active Brownian particle navigating in a quasi-1D environment. However, this analogy may break down in the regime of strong fluctuations.

As we have already shown in Sect. 4.2, this can be quantified by computing the Kullback-Leiber divergence D_{KL} between the distributions obtained from the direct numerical simulations P_{num} and those predicted by the inverse Gaussian law (4.7) with parameters $\langle t \rangle$ and σ extracted from the data. Figure 4.7a shows the corresponding results obtained for all the policies varying the Péclet number. Remarkably, the arrival time distributions of both OP and AAP turn out to be inverse Gaussian for a wide range of noise amplitudes, with significant deviations arising only for $\mathrm{Pe}^{-1} \gtrsim 0.01$. As shown in the heat maps in Fig. 4.7b (left and central panels), when fluctuations are strong the stochastic trajectories are in fact less focused around the optimal path and also more asymmetrically distributed around it. As already discussed in the previous section, this asymmetry leads to larger tails in the probability

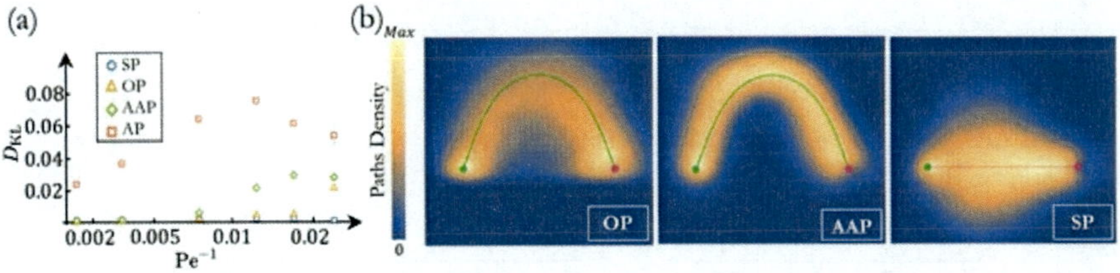

Fig. 4.7 **a** Kullback-Leibler divergence D_{KL} between the numerical and theoretical arrival time distributions as function of the inverse of the Péclet number at $\gamma = 0.7$ (horizontal axis in log scale). **b** Heat maps of 10^3 stochastic trajectories obtained from numerical simulations of OP (left), AAP (center) and SP (right) at Pe $= 40$ and $\gamma = 0.7$. The solid green lines here represent the $D = 0$ Zermelo's path connecting the initial point $\boldsymbol{r}_0 = -\ell/2\hat{\boldsymbol{e}}_x$ and the target $\boldsymbol{r}_{\mathrm{T}} = \boldsymbol{0}$, marked with a green and magenta circle, respectively. The solid red line is instead the straight path linking the two points

distributions (data not shown) and therefore to the observed deviations from the inverse Gaussian law.

In this scenario, the shape of SP distribution turns out to be the most robust to fluctuations. This can be better understood by looking at the corresponding heat map in Fig. 4.7b (rightmost panel). Despite being more dispersed, the stochastic trajectories still look symmetrically distributed around the straight path. This is strictly related to the flow symmetry with respect to the line connecting the starting point with the target (to this end, please refer to Fig. 4.2a).

4.2.3 An Alternative Policy Performance Indicator

Even though in the context of optimal navigation the primary performance indicator is the mean arrival time at the target $\langle t \rangle$, we can define other measures of performances which take into account additional features of the arrival time distribution. Indeed, from our numerical simulations we may get further insights into the performance of a navigation protocol by considering additional observables which also take explicitly into account the effect of fluctuations.

In particular, we shall here consider a new performance indicator which stems from the following remark: owing to the presence of thermal fluctuations, the active particle may reach the target in less time than at $D = 0$. The frequency of these events can be quantified by looking at the probability of arriving before the optimal time t_{opt} in the absence of noise (defined by the Zermelo solution, see Eq. (1.21)): $\text{Prob}(t < t_{\text{opt}}) \equiv \pi_<$. The latter is a measure of how the policies manage to optimize the effect of fluctuations by maximizing the frequency of small arrival time events. Note that since in general $\langle t \rangle \geq t_{\text{opt}}$, $\pi_<$ is bounded by $\frac{1}{2}$ from above.

The new navigation policies introduced in Sect. 4.1 both depend on a free parameter representing the protocol sensitivity. Figure 4.8 shows the policies performances measured from $\langle t \rangle$ and $\pi_<$ as function of sensitivity for both the adaptive aligning policy (AAP, left) and the aligning policy (AP, right). Similarly to the mean arrival

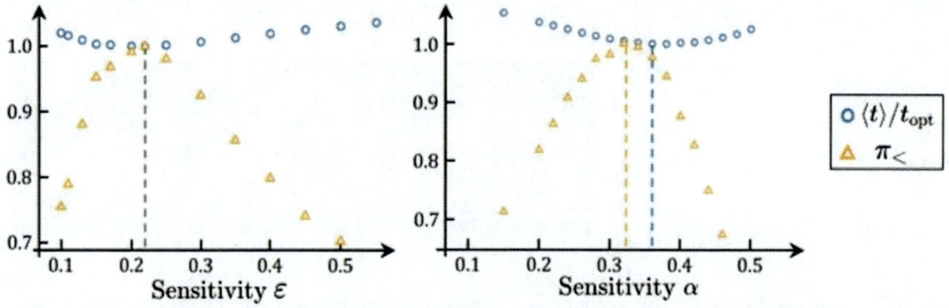

Fig. 4.8 Examples of the curves leading to the determination of the optimal sensitivities ε (AAP, left) and α (AP, right) at $\text{Pe} = 400$ and $\gamma = 0.7$. Blue circles: mean arrival time $\langle t \rangle$ in units of the optimal time in absence of noise t_{opt}. Orange triangles: probability of getting to the target in a time $t < t_{\text{opt}}$. Here, both observables are normalized by their maximum/minimum value. In both plots the corresponding optimal sensitivity values are indicated by the vertical dashed lines

time, the performance indicator $\pi_<$ allows to obtain a clear optimal sensitivity value for both AP and AAP. Moreover, the optimal values obtained independently from $\langle t\rangle$ and $\pi_<$ generally coincide for AAP while some small differences are observed for AP. In the latter case, choosing either of the two estimates does not lead to significant variations of the values of $\langle t\rangle$ and $\pi_<$, and the corresponding arrival time distributions do not differ significantly.

Performance and Robustness Assessment

In Fig. 4.9a we show the probability $\pi_<$ as a function of the relative flow strength γ. On the one hand, the policies performances show trends analogous to those reported in Sects. 4.2, 4.2.2 for the analysis of the mean arrival time, with a preserved hierarchy at strong flows. As expected, in order of increasing performance we also find here SP, AP, AAP and OP. On the other hand, the differences between the various protocols appear more striking. For example, the probability that a swimmer following AAP reaches the target in a shorter time than t_{opt} is around 35% lower than that of a swimmer following OP at $\gamma = 1$, while the corresponding mean arrival times deviate by only a few percent (see Sect. 4.2). As shown in Fig. 4.9b, similar conclusions can

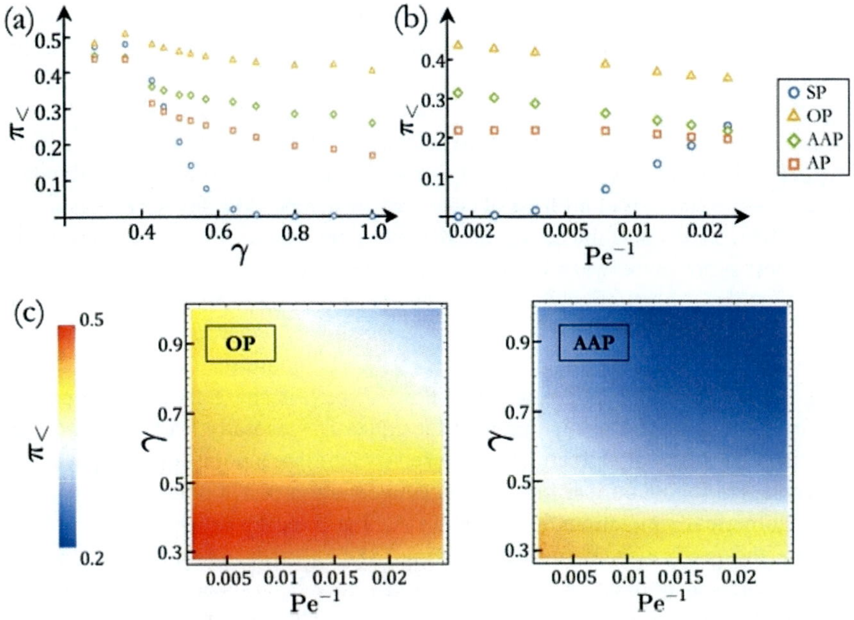

Fig. 4.9 Probability of reaching the target before the optimal time in absence of noise (t_{opt}) as function of **a** the relative flow strength γ at Pe $= 400$ and **b** the inverse of the Péclet number at $\gamma = 0.7$ (horizontal axis in log scale). **c** Colour maps showing the performance of OP (left) and AAP (right) as function of both the relative flow strength γ and the Péclet number Pe. The colour gradient represents the probability of reaching the target faster than in absence of noise from low (blue) to high (red). All data in **a–c** were averaged over 10^5 independent trajectories

be reached examining the behaviour of $\pi_<$ as function of the Péclet number. Namely, all OP, AAP and AP show a slight decrease in performance upon increasing the strength of fluctuations, while SP becomes more advantageous at large noises. For completeness, we also show in Fig. 4.9c the heat maps comparing the values of $\pi_<$ for OP and AAP as function of γ and Pe. Here again, the behaviour of the performance indicator is similar to that of the mean arrival time.

Overall, the new indicator $\pi_<$ is largely dominated by the mean arrival time, although it accounts explicitly for the effect of fluctuations and provides a more refined evaluation of the protocols performances. $\pi_<$ therefore leads to qualitatively analogous conclusions regarding the policies efficiency.

4.3 Optimal Navigation on a Manifold

We next show how AP and AAP navigation protocols are applicable to motion on curved landscapes. As we have already extensively discussed in Ch. 2, self-propelled motion on curved surfaces has indeed recently earned growing attention both at individual [25, 26] and collective [27, 28] levels. As stochastic motion taking place on a generic Riemannian manifold involves multiplicative noise, solving the corresponding MFPT equation (4.2) requires advanced computational techniques [29], which will introduce additional challenges for determining the stochastic optimal control (4.3). On the other hand, we have shown in Chap. 2 that it is possible to generalize Zermelo's approach to self-propelled motion on curved surfaces using a mapping to Finsler geometry [30]. The corresponding noiseless optimal path—Randers geodesic—can then straightforwardly be used to extend AP and AAP policies to non-Euclidean spaces, while in general SP corresponds to the particle pointing along the shortest geodesic between its position and the target (details in Appendix A.1).

For the sake of illustration, let us consider the case of active motion on a sphere in the presence of a unidirectional flow $\boldsymbol{f}(\theta, \phi) = v_f \sin\theta \, \hat{\boldsymbol{e}}_\phi$, where θ and ϕ respectively denote the polar and azimuthal angles in the spherical coordinate system. As shown in Fig. 4.10a (see the black arrows), this flow—which is characterized by a pair of vortices at the poles and is maximum at the equator—generally leads to non-trivial Randers geodesics (solid green line) between two arbitrary points on the sphere.

Simulating the counterpart of the Langevin equation (4.1) on the sphere (details on the numerical simulations in Appendix A.1) at fixed $\mathrm{Pe} = 10^3$, we are able to compare the performances of AP, AAP and SP. As in the 2D case, the optimal values of the sensitivities α (AP) and ε (AAP) have been determined empirically selecting the one minimizing the mean arrival time. The corresponding arrival time distributions are shown in Fig. 4.10b.

As for the Taylor-Green flow in flat space, we find that they are all in good agreement with the inverse Gaussian law (4.7). In the AP case, this result is probably due to the rather large value of Pe chosen for convenience, which allows all policies to exhibit trajectories well distributed around a one dimensional path. Figure 4.10c moreover shows that at small flow strengths all policies perform similarly with $\langle t \rangle \gtrsim$

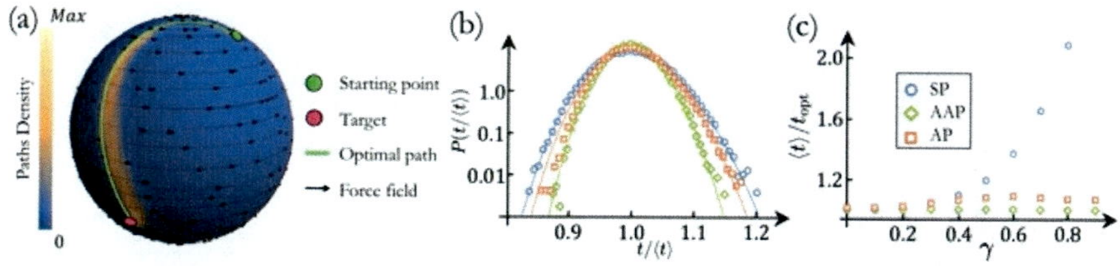

Fig. 4.10 Optimal navigation on a sphere. **a** Illustration of the chosen setup. The black arrows indicate the direction of the flow while the solid green curve connecting the points $\boldsymbol{r}_T = (2\pi/3)\hat{\boldsymbol{e}}_\theta + (3\pi/2)\hat{\boldsymbol{e}}_\phi$ (magenta circle) and $\boldsymbol{r}_0 = (\pi/6)\hat{\boldsymbol{e}}_\theta$ (green circle) corresponds to the Randers geodesic at $\gamma = 0.5$. Here the colour map shows the trajectory distribution for AAP. **b** The arrival time distribution as a function of $t/\langle t\rangle$ for straight (blue circles), Adaptive Aligning (green diamonds) and Aligning (red squares) policies at $\gamma = 0.5$. **c** Mean arrival time as a function of γ for the three policies; the caption is the same as **(b)**. In **a–c** the value of the Péclet number is set to $\mathrm{Pe} = 10^3$. All data in **b**, **c** are averaged over 10^5 independent trajectories

t_{opt}, while for larger γ values leading to more complex Randers geodesics SP becomes increasingly disadvantageous. On the contrary, both AP and AAP always remain close to optimality, as they exploit the information of the noiseless optimal path.

4.4 A New Hierarchy of Navigation Protocols

Even though the new policies introduced so far show remarkable performances as compared to OP, some differences persist. In particular, as shown in Fig. 4.3d, e, while trajectories following OP are symmetrically distributed around Zermelo's path, this is not the case for AP and AAP.

This symmetry can be better appreciated from the OP heat map of stochastic trajectories in the reference frame of the noiseless optimal path, as displayed in Fig. 4.11a. It actually turns out that their distribution in the normal direction with respect to Zermelo's path corresponds to a Gaussian with zero mean. Indeed, as demonstrated by the three different fits in Fig. 4.11b, although the width of this distribution varies at each point along the path, the agreement with the Gaussian always holds. Building on this observation, we now show how it is possible to derive in a semi-analytical fashion a new set of navigation strategies that may lead to further performance improvements.

4.4.1 *Quasi-1D Description of the Problem*

Let us consider the reference frame $\{\hat{\boldsymbol{t}}, \hat{\boldsymbol{n}}\}$ of Zermelo's path and identify the latter with the x-axis without any loss of generality. The Cartesian (x, y) coordinates from now on will therefore correspond to the relative position of the swimmer with

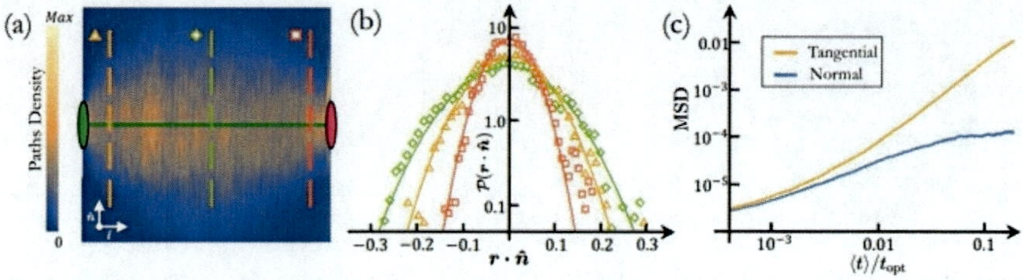

Fig. 4.11 Analysis of the stochastic trajectories in Zermelo's path reference frame. **a** Heat map of stochastic trajectories obtained from numerical simulations of OP in the reference frame $\{\hat{t}, \hat{n}\}$ of Zermelo's path (solid green line). **b** Distributions of the relative distance $r \cdot \hat{n}$ from Zermelo's path at different points as highlighted by the corresponding dashed lines in **a**. Solid curves show the Gaussian fit from the numerical data. **c** Mean square displacement (MSD) over time obtained from averaging over AAP stochastic trajectories. The MSD in the normal direction with respect to Zermelo's path (blue curve) saturates much earlier than the MSD in the tangential one (orange curve), which proves a clear separation of time scales in the system dynamics. The numerical simulations leading to **a–c** have been carried out in the same setup shown in Fig. 4.2a with Pe = 400 and $\gamma = 0.7$. All data in **a–c** are averaged over 10^3 independent trajectories

respect to Zermelo's path in the tangential and normal direction, namely $x = \boldsymbol{r} \cdot \hat{\boldsymbol{t}}$ and $y = \boldsymbol{r} \cdot \hat{\boldsymbol{n}}$.

Neglecting possible effects due to the curvature of such trajectory, we can readily write the Fokker-Planck equation [31] corresponding to the non-dimensionalized version of the overdamped dynamics (4.1):

$$\begin{cases} \partial_t \mathcal{P}(x, y, t) = -\nabla \cdot \boldsymbol{J} \\ \boldsymbol{J} = (\hat{\boldsymbol{u}} + \boldsymbol{f})\mathcal{P}(x, y, t) - \mathrm{Pe}^{-1}\nabla\mathcal{P}(x, y, t)\,, \end{cases} \tag{4.9}$$

where $\mathcal{P}(x, y, t)$ defines the probability of finding the microswimmer at a position (x, y) at time t.

At the same time, all the navigation protocols studied so far in the presence of noise, including OP, essentially amount to making the microswimmer move as fast as possible along Zermelo's path (the x-axis in this simplified scenario) while being confined around it. If the optimal trajectory is long enough, the motion in the transverse direction y could then reach a steady state on a time scale much smaller than the typical navigation time (t_{opt}). This assumption is confirmed by the outcome of our numerical simulations of AAP[2] in the Taylor-Green vortex flow (see setup in Fig. 4.2a). As shown in Fig. 4.11c, the MSD of the microswimmer along the direction normal to Zermelo's path indeed plateaus contrary to the MSD along the tangential direction.

[2] We have also checked that analogous results are obtained considering other policies like AP.

Thanks to this observation, we may thus impose the stationary condition along y in Eq. (4.9), which yields

$$J_y = 0 \Longrightarrow \mathcal{P}(x, y, t) = \mathcal{N}(x, t) \exp\left[\mathrm{Pe} \int_0^y (\hat{u}_y(x, z) + f_y(x, z))\, dz\right], \quad (4.10)$$

where $\mathcal{N}$ is a normalization prefactor that depends in general on both x and t. Earlier we observed that the distribution of trajectories following OP is well-fitted by a Gaussian centered around Zermelo's path. This remark gives us an interesting insight about what OP effectively does and, at the same time, provides a precise indication about how to design a new efficient strategy to be adopted by the microswimmer.

More specifically, if we require the distribution $\mathcal{P}$ in Eq. (4.10) to be locally Gaussian around $y = 0$ for any value of x, we will find the control $\hat{u}_y(x, y)$ shall be given by

$$\hat{u}_y(x, y) = \begin{cases} +1 & \text{if } \; y + \varepsilon f_y \leq -\varepsilon \\ -f_y(x, y) - \frac{y}{\varepsilon} & \text{if } \; |y + \varepsilon f_y| < \varepsilon \\ -1 & \text{if } \; y + \varepsilon f_y \geq \varepsilon \end{cases}, \quad (4.11)$$

where $\varepsilon > 0$ is the protocol sensitivity[3]. This navigation strategy is essentially a generalisation of AAP as also in this case the orientation of the microswimmer is linearly dependent on its distance from Zermelo's path. The novelty lies in the fact that now it also exploits local knowledge of the flow by compensating for its transverse component. It is for these reasons that we will hereafter refer to this strategy as the *Compensating Adaptive Aligning Policy* (CAAP).

The idea of offsetting the transverse flow component unveils a new class of navigation strategies. These can be designed by gradually simplifying the expression (4.11) of CAAP. Indeed, if the microswimmer knows the local flow but not its distance from Zermelo's path, it has two options:

1. Either re-orient itself so as to just compensate for the transverse component of the flow, in which case we will say that the swimmer follows the *Compensating Policy* (CP);
2. Or, in addition to this compensation, keep a fixed angle with respect to Zermelo's path. This could be beneficial since it better confines the microswimmer around it, as in the AP case. This is why we will refer to this strategy as the *Compensating Aligning Policy* (CAP).

Hence, this short analysis has therefore allowed us to define a set of navigation strategies that add to those studied previously. In Table 4.1, we enumerate all of them together with their respective analytical expression of the normal component of the control, i.e. $\hat{u}_y$, and a short description. They can all be seen as part of a hierarchy of

[3] The essential difference between this policy and OP is that while the former is characterized by a constant sensitivity ε that fixes the variance of the (Gaussian) distribution $\mathcal{P}$, in the latter case ε can vary along Zermelo's path. However, the sensitivity cannot be space-dependent since it would break the semi-autonomous nature of the policies.

Table 4.1 Summary of all the protocols we have designed for semi-autonomous optimal navigation together with the corresponding acronyms, analytical expression (if applicable), and short description. They are listed in order of decreasing complexity from OP (top) to SP (bottom)

Name	Acronym	Expression of $\hat{u}_y$	Description
Optimal policy	OP	N/A	Solution from SOC theory
Compensating adaptive aligning policy	CAAP	$-f_y(x, y) - y/\varepsilon$	Compensates for transverse flow and aligns with Zermelo's path depending on the distance
Compensating aligning policy	CAP	$-f_y(x, y) + \cos\alpha$	Compensates for transverse flow and bounds swimmer to Zermelo's path with a constant angle
Adaptive aligning policy	AAP	$-y/\varepsilon$	Aligns with Zermelo's path depending on the distance.
Compensating policy	CP	$-f_y(x, y)$	Compensates for transverse flow
Aligning policy	AP	$\cos\alpha$	Bounds swimmer to Zermelo's path with a constant angle
Straight policy	SP	$\sin\left(\frac{y-y_T}{x-x_T}\right)$	Points straight to the target

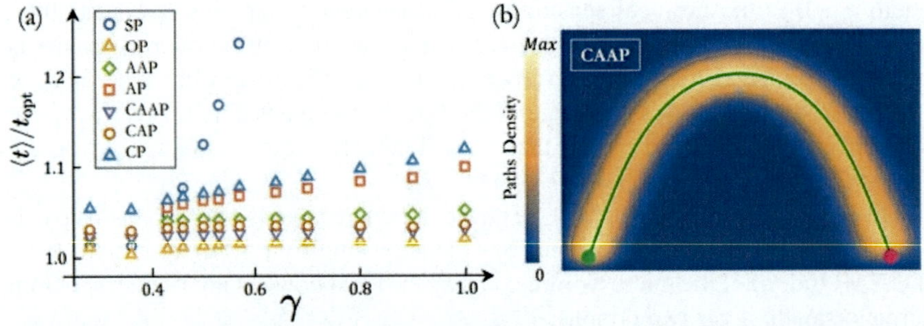

Fig. 4.12 Performance comparison among all the policies in the Taylor-Green flow at Pe = 400. **a** Mean arrival time—normalized by the noiseless optimal time t_{opt}—as a function of the relative flow amplitude γ. All data here are averaged over 10^5 independent trajectories. **b** Heat map of 10^3 stochastic paths obtained from numerical simulations of CAAP. Note the remarkable resemblance with the OP heat map in Fig. 4.3d

navigation strategies listed from top to bottom in order of decreasing complexity—measured in terms of the amount of information required for their implementation.

At this point, we can readily test all these protocols in the same Taylor-Green flow setup as the one illustrated in Fig. 4.2a. We once again quantify their performance by measuring the mean arrival time at the target $\langle t \rangle$ as a function of the relative flow amplitude γ. The corresponding results[4] shown in Fig. 4.12a reveal that, when the

[4] Note that for each policy we always select the sensitivity value (ε or α) minimizing the mean arrival time $\langle t \rangle$.

flow is strong enough ($\gamma \gtrsim 0.4$), the performances of the different strategies generally match their complexity, with CAAP exhibiting strikingly close performances to those of OP (see also the corresponding heat map in Fig. 4.12b). The only exception in this regard is AP, which outperforms CP. This interestingly suggests that being confined around the noiseless optimal path can be more beneficial to the microswimmer than compensating for the transverse flow.

We shall now verify the generality of the results presented here. To this end, we will therefore measure both the performance and robustness of these protocols by implementing them in a Gaussian random flow field.

4.4.2 Semi-autonomous Optimal Navigation in a Gaussian Random Flow

Let us consider a two-dimensional time-independent flow whose velocity field $\boldsymbol{f}(\boldsymbol{r})$ is defined as

$$\boldsymbol{f}(\boldsymbol{r}) = \frac{1}{\sqrt{2}} \nabla \times [\hat{\boldsymbol{e}}_z \psi(\boldsymbol{r})] , \tag{4.12}$$

with $\hat{\boldsymbol{e}}_z$ being the unit-vector in the z-direction and $\psi(\boldsymbol{r})$ a random stream function with zero mean and correlation function [32]

$$\langle \psi(\boldsymbol{r}) \psi(\boldsymbol{r}') \rangle = \ell^2 v_{\mathrm{f}}^2 \exp\left[-\frac{|\boldsymbol{r} - \boldsymbol{r}'|^2}{2\ell^2} \right] , \tag{4.13}$$

where ℓ and v_{f} are the characteristic length and flow intensity scales. The expressions (4.12–4.13) define a so-called Gaussian random flow field. More details on its numerical implementation can be found in Appendix A.2. The corresponding velocity field is shown in Fig. 4.13a. There, it is also displayed the new navigation task, which is to go from the green to the magenta circle in the shortest time in the presence of translational noise.

On the one hand, in order to determine the optimal control map for OP, one would have to solve the corresponding HJB equation (4.2). However, solving such PDE in this complex setup would arguably demand a greater computational effort than finding Zermelo's path, which just requires to solve the corresponding ODE (1.21) by approximating the flow field gradients via finite differences. Then, one simply has to select the proper initial self-propulsion orientation by means of a shooting method. This noiseless optimal solution in our setup is shown in Fig. 4.13a, c as a green curve connecting the initial and final points, and can be used to readily implement all the policies we have designed for semi-autonomous navigation.

The results obtained from our numerical simulations are summarized in the table in Fig. 4.13b. As expected, CAAP is the strategy performing best in terms of mean arrival time at the target, with a significant improvement of ~35% with respect to

(a)

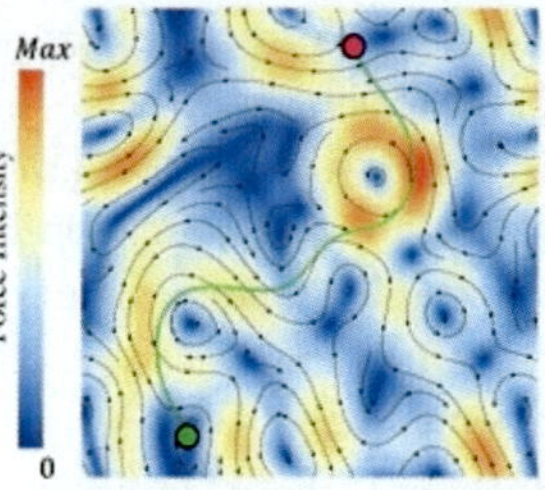

(c)

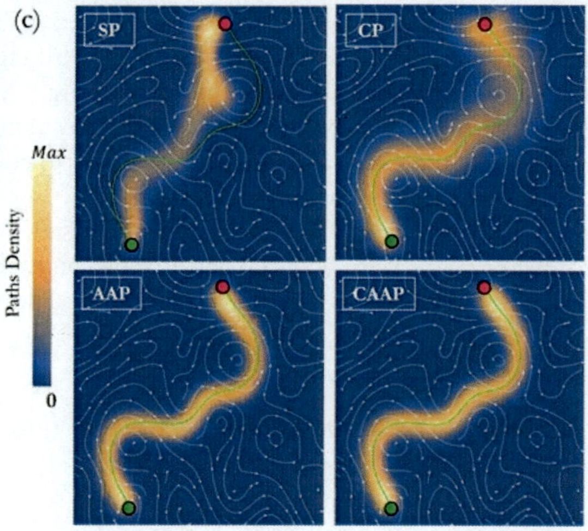

(b)

Policy	$\langle t\rangle/t_{\mathrm{opt}}$	D_{KL}
CAAP	1.028(1)	0.0014(1)
CAP	1.040(1)	0.0034(2)
AAP	1.050(1)	0.0056(4)
CP	1.150(1)	0.22(1)
AP	1.103(1)	0.047(3)
SP	1.574(1)	0.018(1)

Fig. 4.13 Performance comparison among all the policies in the Gaussian random flow. **a** Gaussian random flow map with the black arrows being the flow direction and the colour gradient indicating its intensity. The green curve represents Zermelo's path connecting the starting point $\boldsymbol{r}_0 = 31\ell/6\hat{\boldsymbol{e}}_x + 9\ell/2\hat{\boldsymbol{e}}_y$ and the target $\boldsymbol{r}_{\mathrm{T}} = 25\ell/3\hat{\boldsymbol{e}}_x + 35\ell/3\hat{\boldsymbol{e}}_y$, indicated with a with a green and magenta circle, respectively. **b** Table containing the mean arrival time—normalized by the noiseless optimal time t_{opt}—and the Kullback-Leibler divergence D_{KL} between the numerical and theoretical arrival time distributions. The results reported here have been obtained from data averaged over 10^4 independent trajectories for each policy. The number in parentheses denotes the uncertainty on the last digit. **(c)** Heat maps of 10^3 stochastic paths obtained from numerical simulations of SP, CP, AAP, and CAAP. The white arrows here indicate the flow field direction. Note the robustness of boht AAP and CAAP, which can guide the microswimmer along Zermelo's path (green curve) All the results shown in **a–c** have been obtained from numerical simulations at $\gamma = 0.6$ and Pe $= 60$

the trivial SP. Remarkably, AP and AAP are both still able to perform quite well, showing respectively mean arrival times just $\sim 2\%$ and $\sim 7\%$ higher than the more involved CAAP.

Moreover, the computed Kullback-Leibler divergence D_{KL} between the numerical arrival time distribution and the one predicted by the inverse Gaussian law (4.7) (without fitting parameters) reveals the robustness of the mapping to a 1D driven Brownian motion. Indeed, the distributions of the more sophisticated strategies (CAAP, CAP and AAP) are all still inverse Gaussian even in this complex scenario, with $D_{\mathrm{KL}} \lesssim 6 \cdot 10^{-3}$. In contrast, the more basic AP, SP and especially CP do not really follow this law, with the distance between the numerical and theoretical distributions being one or two orders of magnitude larger.

To this end, it is instructive to look at the heat maps of stochastic trajectories shown in Fig. 4.13c. There, one can indeed see how microswimmers following SP are not well focused around a one-dimensional path since they are likely to get transiently trapped in some local flow structures (e.g. when close to vortices), which also explains the poor performance exhibited by such strategy.

This is in contrast to what happens with the other policies, which manage to avoid such hurdles and thus save time by exploiting the flow via some local knowledge of Zermelo's path. Indeed, the CP heat map in Fig. 4.13c reveals that even the compensation of the transverse flow alone manages to guide the swimmer quite well to the target, yielding a mean arrival time comparable with AP and 27% lower than SP. In the proximity of the arrival point, however, the swimmer following CP may lose time as it is not strongly focused around Zermelo's path. This in turn affects the large time tail of the arrival time distribution and thus causes the deviations from the Inverse Gaussian law remarked above.

Conversely, the paths obtained from numerical simulations of CAAP are symmetrically distributed around Zermelo's path. This is consistent with the increased performance compared to AAP, whose paths are not well centered around the noiseless optimal trajectory, albeit being focused around it.

Overall, the protocols performances once again seem to correspond to the amount of information at their disposal, with the only exception of CP, which is still performing slightly worse than AP because of the greater dispersion of its trajectories. The question, however, is whether the complexity of such strategies is always a synonym for efficiency. In the next section, we will address this point by considering a simple yet instructive setting.

4.4.3 A Toy Model to Explain the Protocols Hierarchy

In order to simplify the problem of optimal navigation, one can observe that typically Zermelo's path consists of making the most of the flow. This is indeed consistent with what we observe in both the Taylor-Green flow setup shown in Fig. 4.2a and the Gaussian random flow one in Fig. 4.13a.

Based on this simple argument, we can thus consider a toy model which is an idealized version of the optimal navigation problem in the two-dimensional plane. Namely, we first neglect the curvature of Zermelo's path and assume it to be a straight line.[5] Then, for it to be optimal, a minimal choice for the flow is given by $\boldsymbol{f}(\boldsymbol{r}) = f_x(y)\hat{\boldsymbol{e}}_x + f_y\hat{\boldsymbol{e}}_y$ whose components are

$$\begin{cases} f_x(y) = \gamma \cos\beta(1 - |y|)\Theta(1 - |y|) \\ f_y = \gamma \sin\beta\,, \end{cases} \tag{4.14}$$

where we have rescaled space and time so as to non-dimensionalize the dynamics of the system (4.1). Here, Θ is the Heaviside step function, while the dynamical regime is fully determined by three dimensionless parameters. Apart from the already introduced Péclet number Pe and relative flow amplitude γ, we indeed have the new parameter β. As illustrated in Fig. 4.14a, it represents the angle between the flow

[5] This approximation implicitly implies a length scale separation between the curvature of the optimal path and the width of the distance distributions of the stochastic trajectories.

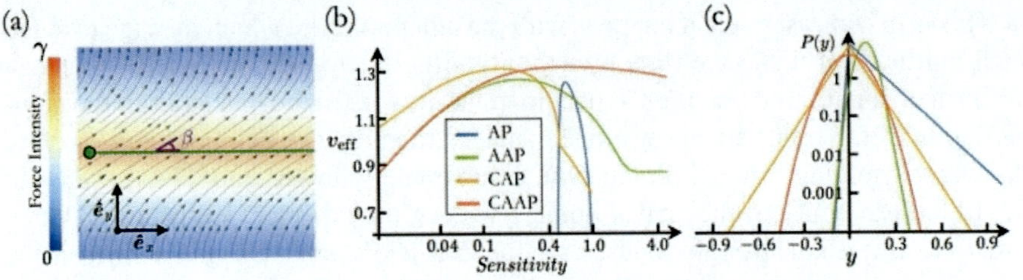

Fig. 4.14 Construction of a simplified model for optimal navigation. **a** Illustration of the flow, with the grey arrows representing its direction and the colour coding its intensity. In this setup, the noiseless optimal path linking any pair of points on the x-axis corresponds to the x-axis itself, as indicated by the green line. The angle β between the flow and the x-axis is here set to $\beta = \pi/4$. **b** Plot of the effective drift $v_{\rm eff}$ along the noiseless optimal path as a function of the protocol sensitivity (α or ε, depending on the policy). All four curves here show a clear maximum, thus providing a well-defined optimal value of this free parameter. **c** Distributions of the distance from Zermelo's path obtained by setting the value of the sensitivity to the one that maximises the effective drift for each strategy. Legend as in **c**. Parameters in **c**, **d** are set to Pe $= 100$, $\gamma = 0.7$, $\beta = \pi/4$

direction and the x-axis. It can thus essentially be tuned so as to change the flow orientation, going from being parallel to $\hat{\boldsymbol{e}}_x$ at $\beta = 0$ (no transverse component) to being exactly orthogonal to it at $\beta = \pi/2$ (no parallel component). Thanks to this flow choice, as long as $\beta \in [0, \pi/2]$ the x-axis indeed corresponds to the noiseless optimal path an active particle shall follow when starting from a point $\boldsymbol{r}_0 = x_0\hat{\boldsymbol{e}}_x$ and aiming for a target placed in $\boldsymbol{r}_{\rm T} = x_T\hat{\boldsymbol{e}}_x$.

Moreover, thanks to the absence of curvature in this simplified setup the correspondence between the swimmer's overdamped dynamics (4.1) and the Fokker-Planck equation (4.9) introduced above is exact. If we now consider the situation in which the target is far from the starting point, i.e. $x_T \gg x_0$, we can then assume a time scale separation between the $\hat{\boldsymbol{e}}_x$ and $\hat{\boldsymbol{e}}_y$ directions and thus safely work under the same assumptions as in Sect. 4.4.1. Now, by imposing the stationary condition along y in Eq. (4.9), i.e. $J_y = 0$, the expression of the probability distribution $\mathcal{P}$ further simplifies into

$$\begin{cases} \mathcal{P}(x, y, t) = Q(x, t)P(y) \\ P(y) = \mathcal{N} \exp\left[\mathrm{Pe} \int_0^y (\hat{u}_y(z) + f_y)\, dz\right] \end{cases}, \tag{4.15}$$

with also $Q(x, t)$ being a properly normalized distribution[6]. However, in unbounded space the distributions corresponding to SP and CP cannot be normalized. For this reason, here we will focus our analysis only on the protocols which confine the particle around Zermelo's path, namely AP, AAP, CAP and CAAP.

[6] This separation of variables is possible thanks to our flow choice. The transverse flow component f_y in (4.14) is indeed constant, making also the flow-compensating protocols (CP, CAP and CAAP) independent of x.

Plugging now (4.15) into the definition of the probability current $\boldsymbol{J}$ in Eq. (4.9) and averaging over y we finally get

$$\partial_t Q(x,t) = -\partial_x[v_{\text{eff}} Q(x,t) - \text{Pe}^{-1}\partial_x Q(x,t)] \,.$$

We have therefore managed to map our problem to a one-dimensional active Brownian motion, where we have identified

$$v_{\text{eff}} \equiv \int_{-\infty}^{+\infty} (\hat{u}_x(y) + f_x(y)) P(y)\, \mathrm{d}y \tag{4.16}$$

with the policy-dependent effective drift along Zermelo's path, while in this simple model the diffusion turns out not to be directly affected by the navigation protocols. Thanks to this 1D mapping, we can then readily argue that the arrival time distribution is an inverse Gaussian[7] (4.7) whose mean is given by $\langle t \rangle = L/v_{\text{eff}}$, with L being the target distance [24]. Consequently, since the mean arrival time $\langle t \rangle$ scales as $\sim 1/v_{\text{eff}}$, a navigation strategy shall perform better the greater its effective drift.

Hence, thanks to this simple toy model we can easily assess the policies performance by measuring the maximum effective drift (4.16) along Zermelo's path. Remarkably, as shown in Fig. 4.14b, there is always a well-defined value of the sensitivity for each strategy that maximizes this quantity. The corresponding distributions $P(y)$ of the distance from the noiseless optimal path are reported in Fig. 4.14c. As expected, while the flow-compensating strategies feature distributions well-centred around $y = 0$, AP and AAP are manifestly asymmetric due to the presence of a transverse flow.

In order to assess their performance in different dynamical regimes, we can first look at how the effective drift v_{eff} varies for each protocol with the relative flow intensity γ or its angle β relative to the noiseless optimal path. The resulting curves are shown in Fig. 4.15a, b and reveal that when the transverse flow is sufficiently strong, i.e. for β and γ large enough, following the simpler AP may lead to a higher effective drift than AAP.

In order to verify this remark and check for other potential hierarchical inversions among the navigation protocols, we shall now explore the whole $\beta - \gamma$ parameter space at fixed Péclet number ($\text{Pe} = 100$)[8]. To this end, we have thus measured the percentage difference between the effective drifts of each pair of protocols and selected the cases where there exists a regime in which a simpler strategy performs better than a more sophisticated one.

The corresponding results obtained are summarized in Fig. 4.15c. From the colour map on the left, it clearly emerges that when the transverse flow is strong enough AP indeed performs better than AAP, with an effective drift up to 16% higher. This

[7] Moreover, this further justifies why were able to observe the same analogy in more general contexts such as, for e.g., the Taylor-Green flow and the Gaussian random flow.

[8] Here, we restrict our analysis to the regime in which the transverse component of the flow is always smaller than the self-propulsion of the active particle, so that all protocols manage to confine it around Zermelo's path.

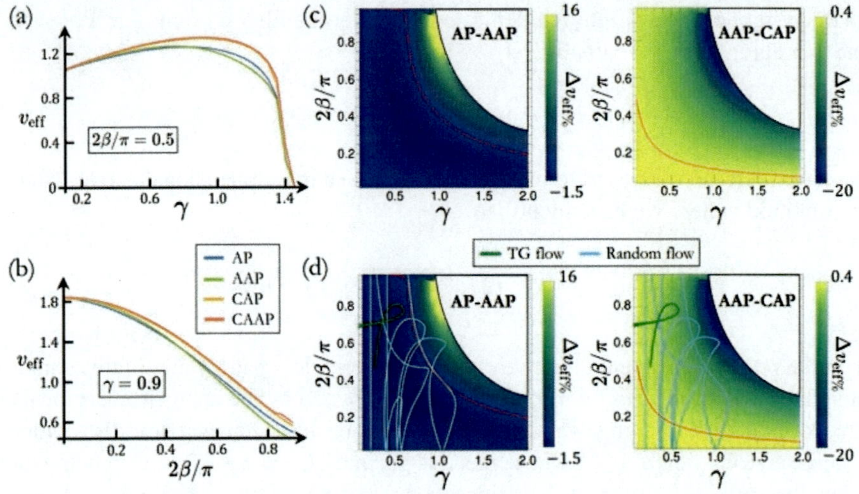

Fig. 4.15 Performance assessment of the navigation policies in different regimes. Plots of the maximum effective drift $v_{\rm eff}$ for each policy as a function of **a** the relative flow amplitude γ at $2\beta/\pi = 0.5$, and of **b** the flow angle with respect to Zermelo's path β at $\gamma = 0.9$. Remarkably, in both cases the two curves corresponding to AP and AAP cross each other at one point: a clear proof of the existence of regimes in which AP may actually perform better than AAP. **c** Colour maps showing the percentage difference between the effective drifts, $\Delta v_{\rm eff\%}$, for two pairs of strategies, namely AP versus AAP (left) and AAP versus CAP (right), as functions of the relative flow amplitude γ and the flow angle with respect to Zermelo's path β in units of $\pi/2$. **d** Same colour maps as in **c** with on top the two curves obtained from the computation of the β and γ values along Zermelo's path in the Taylor-Green flow (green curve, setup and parameters as in Fig. 4.2a) and the Gaussian random flow (cyan curve, see Fig. 4.13a as a reference). The red curves in **c**, **d** denote the points where the difference between the strategies is zero. All data in **a**–**d** are obtained at Pe = 100

effect can be intuitively understood from the distance distributions $P(y)$ shown in Fig. 4.14c. There, one can in fact see how the maximum of the AAP distribution gets shifted due to the transverse flow, while AP manages to keep it in the centre. As a result, in the presence of a strong enough transverse flow, the active particle will be better confined around Zermelo's path if it follows the latter strategy rather than AAP, thus leading to a larger effective drift.

On the contrary, when the transverse flow component is very small, another hierarchical inversion among the protocols may occur: in such a case, AAP is in fact almost equivalent to CAAP and could thus perform better than CAP. However, as shown in the colour map on the right in Fig. 4.15c, the percentage difference between the respective effective drifts is quite marginal (up to 0.4%) such that we cannot actually claim an inversion in the hierarchy in this case as it might even depend on the choice of the flow.

At this point, it is worth discussing which regimes are relevant in actual navigation scenarios. To this end, we have therefore computed the value of γ and β along Zermelo's path in both the Taylor-Green flow setup shown in Fig. 4.2a and the Gaussian random flow one displayed in Fig. 4.13a. The corresponding curves in the $\beta - \gamma$ plane are shown in Fig. 4.15d. It turns out that the flow in the Taylor-Green setup (green curve) does not exhibit much variation in terms of orientation along Zermelo's path, with $0.5 \lesssim \beta \lesssim 0.8$. In contrast, in the random flow case (cyan curve) a much larger region of the parameter space is visited, as one might expect. In either case, however, these curves largely explore just the portion of space in which there is no inversion in the hierarchy of protocols. This is indeed consistent with the performance already measured for each policy via our numerical simulations in the same setups (see results obtained in Sects. 4.2 and 4.4.2).

Lastly, to complete the performance assessment of our navigation policies, we shall also consider their ability to keep the active particle sufficiently close to Zermelo's path as it is crucial for it to eventually hit a finite-size target. In this context, such a measure is provided by

$$\pi_h(\delta) = \int_{-\delta}^{\delta} P(y)\mathrm{d}y \, , \tag{4.17}$$

which is the probability of the active particle hitting a target of size δ located on the x-axis. We have thus computed this quantity for each policy from the distributions obtained with the value of the sensitivity maximizing the effective drift. In Fig. 4.16 are shown two colour maps showing the difference between the hitting probabilities $\Delta\pi_h$ of the same pairs of policies studied above. Remarkably, AP here shows a significantly better performance than AAP in almost the entire parameter space. In contrast, the AAP hitting probability is almost always smaller than that of CAP, showing only a marginal improvement (up to $\Delta\pi_h = 0.04$) when the transverse flow is small. This is actually consistent with the distance distributions $P(y)$ displayed in Fig. 4.14c: the AAP one gets indeed shifted due to the presence of a transverse flow, while both the AP and CAP manage to keep the maximum of the distribution at $y = 0$.

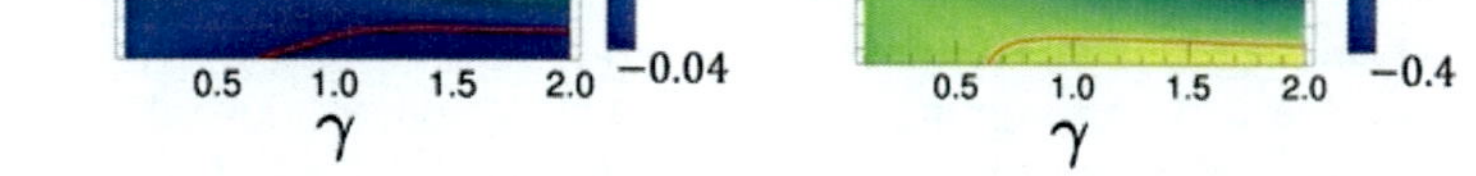

Fig. 4.16 Colour maps representing the difference in hitting probability $\Delta\pi_h$ obtained from two pairs of strategies by selecting the sensitivity which maximizes the effective drift. Namely, here are shown AP versus AAP (left) and AAP versus CAP (right), as functions of the relative flow amplitude γ and the flow angle with respect to Zermelo's path β in units of $\pi/2$. The red curves denote the points where the difference between the strategies is zero. Results here obtained at Pe $= 100$ and with the target size set to $\delta = 0.05\ell$

The results obtained here thus confirm and strengthen the general picture already suggested by the study of the effective drift, i.e. that navigation efficiency is often but not always directly related to the protocol complexity.

4.5 Summary and Discussion

Our analysis has shown how the challenging problem of optimal navigation in the presence of noise can be boiled down to a few simple principles. These essentially aim to stabilize the swimmer motion around the noiseless optimal trajectory, i.e. Zermelo's path.

Starting from the observation of the heat maps of stochastic trajectories obtained with the Optimal Policy (OP) from SOC theory, we indeed first realized how valuable Zermelo's path can be in guiding a swimmer towards the desired target. Based on this remark, we have therefore designed a new class of policies that allow for semi-autonomous optimal navigation of microswimmers in complex and noisy environments. Indeed, the implementation of these policies can be done autonomously by the swimmer using some knowledge of its local environment, such as its relative position with respect to Zermelo's path or the local flow, and then applying simple alignment rules based on such information.

In a first stage, we have focused our attention on strategies that don't make any use of the local flow, namely the Adaptive Aligning Policy (AAP), for which the self-propulsion orientation varies with the distance to Zermelo's path, and the simpler Aligning Policy (AP), in which the swimmer keeps a fixed angle with respect to such path. Despite relying on much simpler principles, they both show performances relatively close to those obtained with OP.

We have then illustrated a way to further improve these results by designing policies based on the ability of some swimmers to adapt their swimming direction according to the local flow field [16, 17]. More specifically, by including the compensation for the transverse component of the flow with respect to Zermelo's path we found an additional increase in the efficiency of our protocols, with the Compensating Adaptive Aligning Policy (a direct extension of AAP) showing performances and heat maps of the stochastic paths almost identical to those obtained from simulations of OP.

In order to test the robustness of these protocols, we have then implemented them in a Gaussian random flow. Remarkably, the newly designed strategies show performances very close to optimality even in such a complex environment, with arrival time distributions matching an inverse Gaussian, i.e. the First Passage Time distribution of a one-dimensional active Brownian motion. Our results thus suggest that the particle confinement around Zermelo's path due to the protocols actually reduces the problem to an effective 1D motion.

Thanks to this insight, we then introduced a toy model for the problem of optimal navigation so as to carry out a more systematic analysis of the protocols performance. In this way we uncovered some dynamic regimes in which, surprisingly enough, the efficiency level of a policy does not correspond to its degree of complexity. In fact, we have shown that, in the presence of a flow with a strong transverse component with respect to Zermelo's path, the Aligning Policy may perform better than the Adaptive Aligning Policy, despite relying on less information.

Moreover, the introduced navigation strategies have the additional advantage of being easily applicable to the problem of optimal navigation on curved surfaces. In an illustrative example on spherical geometry, the semi-autonomous navigation strategies were once again found to perform significantly better than the trivial one consisting of pointing straight to the target.

Overall, our analysis shows that the main factor determining the differences among the policies performance is the presence of a transverse flow, such that the best performing strategies are those that stabilize best the swimmer around Zermelo's path. Then, it is no surprise that the most efficient policies were also found to be the most robust to environmental changes, such as stronger translational diffusion and the introduction of rotational noise.

As a final remark, we note that the constraint of imposing a constant sensitivity (represented by the parameters α and ε) throughout the active particle motion might restrict the performances of the policies. While making these parameters explicitly space dependent would break the semi-autonomous nature of the policies, determining the functions $\alpha(\boldsymbol{r})$ or $\varepsilon(\boldsymbol{r})$ is certainly a simpler problem than calculating the full optimal control map of the swimmer orientation.

References

1. Piro L, Mahault B, Golestanian R (2022) Optimal navigation of microswimmers in complex and noisy environments. New J Phys 24(9):093037. https://doi.org/10.1088/1367-2630/ac9079
2. Bellman R (1954) The theory of dynamic programming. Bull Am Math Soc 60(6):503–515. https://doi.org/10.1090/S0002-9904-1954-09848-8
3. Kappen HJ (2005) Path integrals and symmetry breaking for optimal control theory. IOP Publishing 2005(11):11011
4. Yong J, Zhou XY (1999) Stochastic controls: hamiltonian systems and HJB equations, vol 43. Springer Science & Business Media. https://doi.org/10.1007/978-1-4612-1466-3
5. Muiños-Landin S, Fischer A, Holubec V, Cichos F (2021) Reinforcement learning with artificial microswimmers. Sci Robot 6(52):eabd9285. https://www.science.org/doi/abs/10.1126/scirobotics.abd9285
6. Yang Y, Bevan MA, Li B (2020) Efficient navigation of colloidal robots in an unknown environment via deep reinforcement learning. Adv Intell Syst 2(1):1900106. https://doi.org/10.1002/aisy.201900106
7. Biferale L, Bonaccorso F, Buzzicotti M, Clark Di Leoni P, Gustavsson K (2019) Zermelo's problem: optimal point-to-point navigation in 2d turbulent flows using reinforcement learning. Chaos 29:103138. https://doi.org/10.1063/1.5120370
8. Fernandez-Rodriguez MA, Grillo F, Alvarez L, Rathlef M, Buttinoni I, Volpe G, Isa L (2020) Feedback-controlled active brownian colloids with space-dependent rotational dynamics. Nat Commun 11(1):4223
9. Chen X-Z, Jang B, Ahmed D, Hu C, De Marco C, Hoop M, Mushtaq F, Nelson BJ, Pané S (2018) Small-scale machines driven by external power sources. Adv Mater 30(15):1705061. https://doi.org/10.1002/adma.201705061

10. Bregulla AP, Yang H, Cichos F (2014) Stochastic localization of microswimmers by photon nudging. ACS Nano 8(7):6542–6550. https://doi.org/10.1021/nn501568e
11. Gomez-Solano JR, Samin S, Lozano C, Ruedas-Batuecas P, van Roij R, Bechinger C (2017) Tuning the motility and directionality of self-propelled colloids. Sci Rep 7(1):14891
12. Tsang ACH, Demir E, Ding Y, Pak OS (2020) Roads to smart artificial microswimmers. Adv Intell Syst 2(8):1900137. https://doi.org/10.1002/aisy.201900137
13. Bennett RR, Golestanian R (2015) A steering mechanism for phototaxis in chlamydomonas. J R Soc Interface 12(104):20141164
14. Dai B, Wang J, Xiong Z, Zhan X, Dai W, Li C-C, Feng S-P, Tang J (2016) Programmable artificial phototactic microswimmer. Nat Nanotechnol 11(12):1087–1092. https://doi.org/10.1038/nnano.2016.187
15. Hong Y, Blackman NMK, Kopp ND, Sen A, Velegol D (2007) Chemotaxis of nonbiological colloidal rods. Phys Rev Lett 99:178103. https://doi.org/10.1103/PhysRevLett.99.178103
16. Saha S, Golestanian R, Ramaswamy S (2014) Clusters, asters, and collective oscillations in chemotactic colloids. Phys Rev E 89:062316. https://doi.org/10.1103/PhysRevE.89.062316
17. Palacci J, Sacanna S, Abramian A, Barral J, Hanson K, Grosberg AY, Pine DJ, Chaikin PM (2015) Artificial rheotaxis. Sci Adv 1(4):e1400214. https://doi.org/10.1126/sciadv.1400214
18. Lagzi I, Soh S, Wesson PJ, Browne KP, and Grzybowski BA (2010) Maze solving by chemotactic droplets. J Am Chem Soc 132(4):1198–1199. https://doi.org/10.1021/ja9076793
19. Zheng J, Dai B, Wang J, Xiong Z, Yang Y, Liu J, Zhan X, Wan Z, Tang J (2017) Orthogonal navigation of multiple visible-light-driven artificial microswimmers. Nat Commun 8(1):1438
20. Frangipane G, Dell'Arciprete D, Petracchini S, Maggi C, Saglimbeni F, Bianchi S, Vizsnyiczai G, Bernardini ML, Di Leonardo R (2018) Dynamic density shaping of photokinetic *E. coli*. eLife 7:e36608. ISSN 2050-084X. 10.7554/eLife.36608. https://doi.org/10.7554/eLife.36608
21. Shneydor NA (1998) Missile guidance and pursuit: kinematics, dynamics and control. Elsevier
22. Galloway KS, Justh EW, Krishnaprasad PS (2013) Symmetry and reduction in collectives: cyclic pursuit strategies. Proc R Soc A: Math Phys Eng Sci 469(2158):20130264
23. Edwards C, Spurgeon S (1998) Sliding mode control: theory and applications, 1st edn. CRC Press, London
24. Folks JL, Chhikara RS (1978) The inverse gaussian distribution and its statistical application–a review. J R Stat Society Ser B (Methodol) 40(3):263–289. ISSN 00359246. http://www.jstor.org/stable/2984691
25. Fily Y, Baskaran A, Hagan MF (2016) Active particles on curved surfaces
26. Apaza L, Sandoval M (2018) Active matter on riemannian manifolds. Soft Matter 14:9928–9936. https://doi.org/10.1039/C8SM01034J
27. Sknepnek R, Henkes S (2015) Active swarms on a sphere. Phys Rev E 91:022306. https://doi.org/10.1103/PhysRevE.91.022306
28. Napoli G, Turzi S (2020) Spontaneous helical flows in active nematics lying on a cylindrical surface. Phys Rev E 101:022701. https://doi.org/10.1103/PhysRevE.101.022701
29. Dziuk G, Elliott CM (2013) Finite element methods for surface pdes. Acta Numer 22:289–396. https://doi.org/10.1017/S0962492913000056
30. Piro L, Tang E, Golestanian R (2021) Optimal navigation strategies for microswimmers on curved manifolds. Phys Rev Res 3:023125
31. Risken H (1996) The Fokker-Planck equation. Springer, Berlin
32. Gustavsson K, Mehlig B (2016) Statistical models for spatial patterns of heavy particles in turbulence. Adv Phys 65(1):1–57. https://doi.org/10.1080/00018732.2016.1164490

Chapter 5
Towards a More Realistic Description of Optimal Navigation

It can scarcely be denied that the supreme goal of all theory is to make the irreducible basic elements as simple and as few as possible without having to surrender the adequate representation of a single datum of experience.

Albert Einstein

So far, the problem of optimal navigation has essentially been a synonym for minimization of the time it takes to explore the surroundings or to reach a target. However, when thinking in practical terms about the efficiency of a microswimmer's motion, it is natural to ask how much energy it consumes to accomplish a given task [1, 2].

Using tools from OC theory, we therefore aim here to extend Zermelo's classical solution and determine how the optimal strategy changes when considering more complex cost functions, i.e. which also include the energetic costs associated with navigation. This will be achieved by relaxing some assumptions of our model.

Firstly, we will study how the optimal navigation policy is modified when allowing the swimmer to control its own self-propulsion speed as well as its heading direction. On the one hand, lifting the constraint on the speed indeed naturally introduces a possible trade-off between the minimization of translational drag dissipation and time. On the other hand, tuning its own self-propulsion may help the swimmer to efficiently explore strong flow regions and potentially reduce the chaotic character of optimal trajectories, as already discussed in Chap. 3.

Furthermore, in our description of the problem we have so far only considered the effect of the external flow in the translational dynamics of the swimmer, neglecting possible torques induced by its presence. This has been possible by approximating the active particle to a point-like object. However, in practice we aim to describe entities, albeit microscopic, with a well-defined size and shape.

Therefore, in order to move towards a more realistic description of microswimmers navigation, we here introduce this ingredient into our model. Specifically, to investigate the direct effect that shape has on motion efficiency, we consider microswimmers moving at constant speed without any active control over their

L. Piro, *Optimal Navigation in Active Matter*, Springer Theses,
https://doi.org/10.1007/978-3-031-52577-3_5

motion, hence referred to as *dumb* swimmers. Starting from the similarities between the angular dynamics obtained from Zermelo's solution and the hydrodynamic torque exerted on an spheroidal body with a disk-like shape, we reveal the existence of a natural trade-off between energy dissipation and time depending solely on the swimmer shape. By studying the exemplary case of self-propelling spheroids in a shear flow, we indeed show it is always possible to identify an optimal shape minimizing the overall cost.

As a final step, we then make these swimmers *smart* by allowing them to actively steer. However, in doing so they inevitably consume more energy, thus introducing an extra cost in the navigation problem. In this context, we thus determine a new set of equations to be solved in order to find the theoretically optimal strategy based on OC theory. We then study its performance while varying the swimmer shape and show the relative variation in the total navigation cost with respect to the *dumb* swimming case. Finally, we also highlight the substantial differences between such smart steering strategy and Zermelo's protocol implemented on swimmers with a well-defined shape.

5.1 Optimal Navigation of an Active Particle with Varying Speed

As a first extension of Zermelo's optimal navigation problem introduced in Sect. 1.3, let us consider an overdamped active particle which can tune both its own self-propulsion speed and direction while moving in the presence of a stationary flow. Accordingly, the equations governing its motion are

$$\dot{\boldsymbol{r}} = \boldsymbol{v} + \boldsymbol{f}(\boldsymbol{r}), \tag{5.1}$$

where $\boldsymbol{v} = v\hat{\boldsymbol{u}}$ with $\hat{\boldsymbol{u}}$ being the particle orientation. Since we are now assuming to have also full control over the self-propulsion speed v, we can extend the navigation problem by aiming at minimizing not only the time of arrival at a target, but also the energy required to reach it.[1]

We shall therefore change our formulation of the optimal navigation problem and consider the following cost function:

$$\mathcal{C}_{\text{th}} = \int_0^{t_f} \left(\frac{\gamma v^2}{2} + \beta \right) dt \tag{5.2}$$

where $\gamma/2$ stands for the translational drag coefficient of the active particle. This cost represents the sum of two contributions, namely the total energy consumed by the particle and its travel time, with the free parameter β acting as a weight between

[1] Besides, if in this context we only considered time minimization, we would obtain the trivial solution in which the particle moves in a straight line to the target with infinite speed.

these two quantities. Moreover, such parameter has the dimension of a power and can in fact be loosely interpreted as the *acceptable* power supplied by the particle to navigate. In any case, this allows us to define the system Hamiltonian (1.9), which takes the form: $\mathcal{H} = \beta + \gamma v^2/2 + \boldsymbol{p} \cdot [\boldsymbol{v} + \boldsymbol{f}(\boldsymbol{r})]$, where the state variable is simply the particle position $\boldsymbol{r}$ and the control corresponds to the self-propulsion vector $\boldsymbol{v}$. In order to minimize the cost (5.2), we shall now apply Pontryagin's principle (1.10) from OC theory:

$$\begin{cases} \dot{p}_i = -p_j \partial_i f_j \\ \frac{\partial \mathcal{H}}{\partial v_i} = 0 \Longrightarrow v_i = -p_i/\gamma \\ \mathcal{H} = \beta + \gamma v^2/2 + \boldsymbol{p} \cdot [\boldsymbol{v} + \boldsymbol{f}(\boldsymbol{r})] = 0 \, , \end{cases} \tag{5.3}$$

where $\partial_i \equiv \partial_{r_i}$ and the last condition, valid at any time since the ODE system is autonomous, determines unambiguously the final time t_{f}. We can then use the second equation in (5.3) to replace $\boldsymbol{p}$ by $\boldsymbol{v}$ in the first one and straightforwardly obtain the dynamical equation for our control variable

$$\dot{v}_i = -v_j \partial_i f_j \, , \tag{5.4}$$

which should be solved together with the equation of motion (5.1) and the boundary conditions $\boldsymbol{r}(0) = \boldsymbol{r}_0, \boldsymbol{r}(t_{\mathrm{f}}) = \boldsymbol{r}_{\mathrm{T}}$, so as to find the trajectories that minimize the total cost (5.2).

Interestingly, since $\dot{v}_i = \dot{v}\hat{u}_i + v\dot{\hat{u}}_i$ we can write the equations for the optimal control of the self-propulsion speed and steering separately

$$\begin{cases} \dot{v} = -v \, \hat{\boldsymbol{u}}^T \mathbb{F} \hat{\boldsymbol{u}} \\ \dot{\hat{\boldsymbol{u}}} = \mathbb{W}\hat{\boldsymbol{u}} - \mathbb{S}\hat{\boldsymbol{u}} + (\hat{\boldsymbol{u}}^T \mathbb{S}\hat{\boldsymbol{u}})\hat{\boldsymbol{u}} \, , \end{cases} \tag{5.5}$$

where we have decomposed the strain rate tensor $\mathbb{F}$ of the external flow field $\boldsymbol{f}$ –with components $F_{ij} = \partial_j f_i$– into its symmetric and anti-symmetric parts as $\mathbb{F} = \mathbb{W} + \mathbb{S}$, with $S_{ij} = (\partial_j f_i + \partial_i f_j)/2$ being the rate of stretching and shearing while $W_{ij} = (\partial_j f_i - \partial_i f_j)/2$ is the rate of rotation. As a first insight from this new navigation strategy, we could note that the right hand side of the first equation can also be written as $-v\hat{\boldsymbol{u}} \cdot \nabla(\boldsymbol{f} \cdot \hat{\boldsymbol{u}})$. Thus, if $\boldsymbol{f} \cdot \hat{\boldsymbol{u}} > 0$, then the speed will increase (respectively decrease) if $\hat{\boldsymbol{u}}$ points down (respectively up) the gradient. On the other hand, if $\boldsymbol{f} \cdot \hat{\boldsymbol{u}} < 0$, we naturally get the opposite and the particle will accelerate until it overcomes a barrier. It is also worth noting that the resulting orientation dynamics in (5.5) is independent of the particle's speed and, even more remarkably, since the expression of the active steering $\dot{\hat{\boldsymbol{u}}}$ corresponds to Zermelo's solution (1.21) in any dimension, we find that the optimal steering policy is not affected by allowing the self-propulsion speed to vary.

5.1.1 Analytical Solution in a Shear Flow

We now want to show the implementation of this smart throttling policy in a simple scenario. To this end, we can consider the same setup used to illustrate Zermelo's protocol in Sect. 1.3, in which motion takes place in a 2D shear flow $\boldsymbol{f}(\boldsymbol{r}) = \kappa y\,\hat{\boldsymbol{x}}$, with $\kappa > 0$ being the shear rate. The active particle starts from the origin and has to reach another point $\boldsymbol{r}_{\mathrm{T}}$ at distance ℓ along the $x-$axis, i.e. $\boldsymbol{r}_{\mathrm{T}} = \ell\,\hat{\boldsymbol{x}}$ with $\ell > 0$, while minimizing the total cost (5.2). According to the solution we have obtained from OC theory (5.4), we should solve the following set of ordinary differential equations

$$\begin{cases} \dot{x} = v_x + y \\ \dot{y} = v_y \\ \dot{v}_x = 0 \\ \dot{v}_y = -v_x\,. \end{cases} \tag{5.6}$$

where we have rescaled space and time as $\boldsymbol{r} \to \ell \boldsymbol{r}$ and $t \to t/\kappa$, such that the velocity is in units of $\kappa\ell$ and we will deal only with non-dimensional quantities. We can then analytically solve this system and get the parametric equations for the optimal trajectory:

$$\begin{cases} x(t) = -\frac{v_{x0}}{6}t^3 + \frac{v_{y0}}{2}t^2 + (y_0 + v_{x0})t + x_0 \\ y(t) = -\frac{v_{x0}}{2}t^2 + v_{y0}t + y_0 \\ v_x(t) = v_{x0} \\ v_y(t) = -v_{x0}t + v_{y0}\,. \end{cases} \tag{5.7}$$

where v_{x0} and v_{y0} are the initial components of the particle self-propulsion and the vector $\boldsymbol{r}_0 = x_0\hat{\boldsymbol{x}} + y_0\hat{\boldsymbol{y}}$ denotes its initial position. Imposing now the boundary conditions mentioned above, we get the expressions of the initial velocity components as function of the unknown arrival time t_{f}:

$$\begin{cases} v_{x0}(t_{\mathrm{f}}) = \frac{12}{12t_{\mathrm{f}}+t_{\mathrm{f}}^3} \\ v_{y0}(t_{\mathrm{f}}) = \frac{6}{12+t_{\mathrm{f}}^2}\,. \end{cases} \tag{5.8}$$

Furthermore, the explicit expression for the total energy dissipated by the swimmer (in units of $\gamma\kappa\ell^2$) can also be easily obtained from Eqs. (5.7–5.8) as function of t_{f}:

$$\mathcal{E} = \frac{1}{2}\int_0^{t_{\mathrm{f}}} v^2\,dt = \frac{6}{12t_{\mathrm{f}} + t_{\mathrm{f}}^3}\,. \tag{5.9}$$

The resulting trajectories will therefore minimize the total cost[2] given by $\mathcal{C}_{\mathrm{th}} = \mathcal{E} + \rho t_{\mathrm{f}}$, where we have defined a new dimensionless parameter that determines the whole system dynamics, namely $\rho \equiv \frac{\beta}{\gamma(\kappa\ell)^2}$.

[2] Here written in units of $\gamma\kappa\ell^2$.

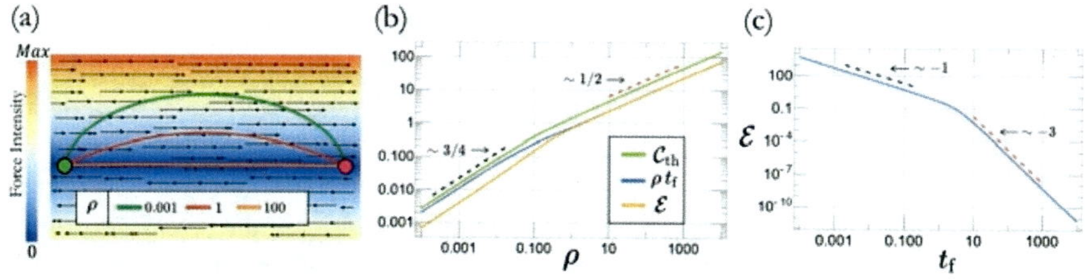

Fig. 5.1 Smart throttling in a shear flow. **a** Three examples of optimal trajectories for different values of ρ. The goal is to reach the target $\boldsymbol{r}_T = \hat{\boldsymbol{x}}$ (magenta circle) starting from the origin (green circle) while minimizing the total cost $\mathcal{C}_{th}$. Large values of this parameter correspond to almost straight paths (see, for e.g., orange curve here for $\rho = 100$), while small values of ρ result in more curved shapes since the particle makes greater use of the flow. The black arrows here indicate the flow direction, while the colour gradient stands for its intensity. **b** Plot as function of ρ of the energetic (orange curve) and time costs (blue curve) that make up the total cost function $\mathcal{C}_{th}$ (green curve) that is minimized by the smart throttling policy. Note the two predicted scaling exponents $\sim 1/2$ (red dashing) and $\sim 3/4$ (black dashing) respectively for large and small values of ρ. **c** Curve showing the energy dissipation $\mathcal{E}$ as function of the arrival time at the target t_f. The optimal combination of the two costs depends on ρ, such that large values of this parameter would correspond to the upper-left branch of this curve and vice versa. The two scaling exponents ~ -1 (black dashing) and ~ -3 (red dashing) here simply reflect the analytical expression of the energy (5.9) derived in this setup

In order to close the system of equations and solve the problem, we just have to determine the travel time t_f. This can be achieved using the condition $\mathcal{H} = 0$ at time $t = 0$. In this case, such expression greatly simplifies since the starting position $\boldsymbol{r}_0$ of the particle is a stationary point of the force field and therefore reads

$$\frac{\mathcal{H}}{\gamma(\kappa\ell)^2} = -v_0^2/2 + \rho = 0 \Longrightarrow v_{x0}^2(t_f) + v_{y0}^2(t_f) = 2\rho\,. \tag{5.10}$$

We have thus obtained an implicit equation to be solved for t_f which, together with Eqs. (5.7–5.9), makes up the analytical solution of the problem.

The optimal paths ultimately depend just on the choice of ρ, which basically stands for the relative weight between energy consumption and travel time, such that the greater ρ, the greater the cost of time and vice versa. This can be readily understood by looking at the trajectories shown in Fig. 5.1a corresponding at three different values of ρ. Their shape goes from being almost straight at large ρ to a more and more curved one as ρ decreases, since then the particle is trying to make a greater use of the flow in order to consume less energy at the expense of taking longer to reach the target.[3]

More specifically, we can distinguish two different regimes while varying ρ. When $\rho \ll 1$, from Eqs. (5.8, 5.10) we may observe that the vertical component of the initial velocity v_{y0} will dominate, such that the arrival time t_f will scale as $\rho^{-1/4}$. This regime essentially amounts to large t_f, which is consistent with our original guess: for small

[3] Moreover, this is consistent with the physical interpretation of ρ that we can gain from Eq. (5.10). It indeed corresponds to the square of the initial speed of the active particle (in units of $\kappa\ell$), such that for small ρ it will make more use of the flow and vice versa.

values of ρ the particle will indeed try to preferably minimize the energy consumption by exploiting more the flow. As a result, by also considering Eq. (5.9), we can predict that both the energy $\mathcal{E}$ and the time contribution ρt_f to the total cost $\mathcal{C}_{th}$ will scale as $\rho^{3/4}$ in this regime.

Conversely, when $\rho \gg 1$, it is clear from Eqs. (5.8, 5.10) that the horizontal component of the velocity v_{x0} will prevail instead, and in this case the arrival time will scale as $\rho^{-1/2}$. This tells us that, indeed, when ρ becomes large the active particle will favour motion in the horizontal direction, therefore pointing more and more straight to the target. In this regime, using the same dimensional argument as above, it can be easily shown that both components $\mathcal{E}$ and ρt_f of the total cost $\mathcal{C}_{th}$ scale as $\rho^{1/2}$.

These two scaling regimes can be clearly observed in Fig. 5.1b. There, we have plotted the total navigation cost $\mathcal{C}_{th}$ as function of ρ, showing separately also both the energetic and time costs which, despite featuring the same trend, differ in magnitude in the two regimes. This is especially noticeable at small values of ρ where it is the energy component $\mathcal{E}$ that is minimized the most.

We can achieve the same kind of insight by looking at the plot of the energy consumption $\mathcal{E}$ as function of the travel time t_f shown in Fig. 5.1c. In this representation, the optimal balance between energy and time is a point on this curve that depends on the choice of ρ. Namely, for $\rho \gg 1$ such point will move to the left, with the active particle primarily minimizing the arrival time, and the energy thus scaling as $\mathcal{E} \sim v^2 t_f \sim t_f^{-1}$. On the contrary, for $\rho \ll 1$ the optimal point will shift to the right along this curve, such that the particle will take longer to reach the target while saving energy more efficiently. This is reflected in the observed scaling $\mathcal{E} \sim t_f^{-3}$, which indeed points out that the self-propulsion speed is no longer inversely proportional to the travel time.

Allowing for smart throttling, we have therefore gained a first insight into a more general version of the optimal navigation problem, where we have considered the energetic cost associated with active motion. However, in our discussion of the problem so far, we have treated the drag coefficient γ as a parameter when, in fact, it is set by the shape of the particle. Moving towards a more realistic description of optimal navigation for microswimmers, we shall now therefore take into account agents with a well-defined size and shape, which also implies to consider the effect of the flow-induced torques on the swimmer dynamics. Since these can have an effect analogous to Zermelo's steering policy, one can observe non-trivial behaviours even for microswimmers that do not apply any active control over their motion.

5.2 Smart Swimming by Dumb Swimmers: The Effect of Shape

5.2.1 Solid Bodies in a Fluid Flow at Low Reynolds Number

In the overdamped limit, the orientation dynamics of axisymmetric microswimmers immersed in a fluid flow can be described by [3]

$$\dot{\hat{\boldsymbol{u}}} = \mathbb{W}\hat{\boldsymbol{u}} + \alpha[\mathbb{S}\hat{\boldsymbol{u}} - (\hat{\boldsymbol{u}}^T\mathbb{S}\hat{\boldsymbol{u}})\hat{\boldsymbol{u}}] \equiv \omega_f(\boldsymbol{r}, \hat{\boldsymbol{u}})\,, \tag{5.11}$$

where $\mathbb{W}$ and $\mathbb{S}$ are flow rotation and strain tensors, respectively. Thanks to symmetry arguments, the angular dynamics of these bodies of revolution depends just on a single parameter α, known as *Bretherton's constant*, which is intrinsically related to the swimmers shape.

An effective yet simple way to study the direct effect of shape on navigation efficiency is to consider spheroidal particles [4], whose Bretherton's constant α is known:

$$\alpha = \frac{\lambda^2 - 1}{\lambda^2 + 1}\,, \tag{5.12}$$

where the particle's aspect ratio is here indicated as $\lambda \equiv b/a$, with b and a being the semi-axes of the spheroid along the self-propulsion direction and the equator, respectively. As a result, negative values of the shape parameter α translate into particles with $a > b$, thus more flattened at the poles and known as *oblate* spheroids, positive values of α instead correspond to more elongated particles known as *prolate* spheroids, while $\alpha = 0$ simply represents a sphere. In this context, such parameter can only assume values in the interval $(-1; +1)$, with the lower and upper bounds respectively achieved in the limits $\lambda \to 0$ (flat disk) and $\lambda \to \infty$ (infinite needle-like particle). Moreover, when specifically discussing the motion of spheroidal bodies in a flow, their angular dynamics (5.11) is actually known as Jeffery's equation [5, 6].

Remarkably, Zermelo's solution for the orientation dynamics in any dimension, corresponding to the second equation in system (5.5), is equal to the Jeffery angular velocity $\omega_f(\boldsymbol{r}, \hat{\boldsymbol{u}})$ experienced by a spheroid with shape parameter $\alpha = -1$, i.e. by a flat disk self-propelling along its own axis of symmetry (in the case of spheroids). Thanks to this exact correspondence, we have therefore found out that a microswimmer with a disk-like shape moving in the presence of any stationary flow will always autonomously follow the time-optimal path without having to actively steer.

However, we shall also take into account that the shape of a swimmer directly impacts the energy it dissipates during its motion. This information is essentially encoded in both the translational and rotational drag coefficients, respectively γ_t and γ_r, of the swimmer,[4] whose values also depend on the boundary conditions on its surface.

In this regard, one could consider the so-called *no-slip* boundary conditions for which the fluid velocity on the surface is exactly zero. This is indeed the case for microswimmers being propelled by external forces, e.g. electromagnetic fields [7, 8], and the corresponding drag coefficients γ_{ns} can be computed exactly for spheroids [9, 10]. In the opposite limit, we would instead have a body whose shear stress at the boundary is zero and the fluid velocity profile is therefore unaltered by the surface. These boundary conditions are called *perfect-slip* and in such case it is still possible

[4] There are actually two pairs of drag coefficients for spheroids (for translations and rotations along and transverse to the axis of symmetry). However, since here the swimmer is propelling along its axis of symmetry and is assumed to not rotate around it, there are only two coefficients that enter the total drag dissipation.

to compute the drag coefficients $\gamma_{\rm ps}$ for spheroidal swimmers using some analytical approximations [11].

However, the real boundary conditions most often lie in between the two limiting cases just described. Indeed, a common self-propulsion mechanism of biological microorganisms is the coordinate beating motion of cilia [12, 13], which effectively generate a slip velocity profile along the surface. Moreover, artificial active particles, e.g. Janus particles [14, 15], rely on self-diffusiophoresis and are thus intrinsically driven by a slip velocity.

Hence, to keep our discussion as general as possible, here we quantify the drag power dissipated by a microswimmer based on the recent work by Nasouri et al. [16], in which the authors define a lower bound for the energy dissipated by a surface-driven swimmer of any shape. Remarkably, the resulting drag coefficients γ_i (with $i = t$ or r) are a simple combination of the ones obtained with no-slip $\gamma_{{\rm ns}_i}$ and perfect-slip $\gamma_{{\rm ps}_i}$ boundary conditions, namely:

$$\gamma_i = \frac{\gamma_{{\rm ns}_i}\gamma_{{\rm ps}_i}}{\gamma_{{\rm ns}_i} - \gamma_{{\rm ps}_i}}, \tag{5.13}$$

which is the expression we are going to use throughout this chapter to compute the drag coefficients of spheroidal swimmers. Here, we will specifically make use of the no-slip $\gamma_{{\rm ns}_i}$ and perfect-slip $\gamma_{{\rm ps}_i}$ drag coefficients given in Refs. [9] and [11], respectively.

Before discussing more in details their dependence on the spheroid aspect ratio, it is worth remarking that such coefficients are proportional[5] to the particle size s. More specifically, the translational drag coefficient γ_t depends linearly on s, while $\gamma_r \sim s^3$, so that for small swimmers we can expect the translational dissipation to dominate the overall cost of navigation. Nevertheless, in the following we will consider a regime in which the swimmer size is not too small so that the cost of active steering can potentially be comparable with the translational drag.

Finally, we can quantitatively show in Fig. 5.2 the relation between the drag coefficients of spheroidal particles and their shape. As just discussed, the translational drag coefficient appears to be typically orders of magnitude larger than the corresponding rotational one, due to the different size prefactor. However, thanks to our choice of the swimmer size, the two become comparable for needle-like swimmers ($\alpha \to 1$), possibly leading to interesting trade-offs in the shape when an active steering is in place (see results in Sect. 5.3). Moreover, γ_r as function of the shape parameter α does not show a monotonic trend, being maximum for the two extreme cases of a disk- or needle-like particle, such that more spherical shapes turn out to be the most favourable in terms of rotational cost. Indeed, a perfectly spherical body ($\alpha = 0$) has null cost ($\gamma_r = 0$) associated to active re-orientation.

Furthermore, on the one hand, by simply looking at the translational drag coefficient γ_t, a disk-like microswimmer ($\alpha \to -1$) turns out to dissipate a huge amount

[5] They are also proportional to the particle mobility. However, hereafter this is set equal to one without loss of generality.

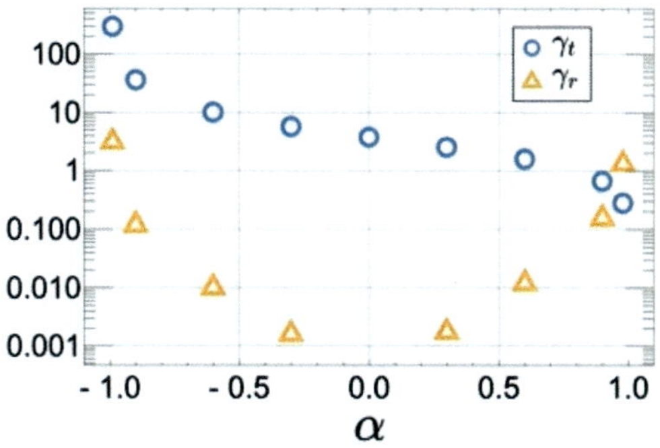

Fig. 5.2 Translational drag coefficient γ_t (blue circles) and rotational drag coefficient γ_r (orange triangles) for spheroids obtained from Eq. (5.13) as function of the shape parameter α. Here, the volume of the particle is kept constant and equal to $V = \frac{4\pi}{3}s^3$, with $s = 0.1$

of energy when compared to more elongated shapes ($\alpha \to 1$), which can save up to $\sim 10^3$ times more (for the same amount of time and same speed). On the other hand, as a result of the aforementioned analogy with Zermelo's solution, a needle-like particle that does not actively steer would inevitably take longer to reach any target, such that there can be regimes in which the optimal shape is somewhere in between a disk and a needle. This emergent trade-off may thus select a non-trivial optimal shape even for microswimmers moving in a fluid flow that do not apply any active control over their own dynamics.

5.2.2 *Spheroids Efficiency in a Shear Flow*

We now aim to study more in details the motion efficiency of spheroidal microswimmers. More specifically, we shall here go back to the case in which the self-propulsion speed is fixed and then compare the overall cost to reach a specific target among particles with different aspect ratios. In this way, the swimmer is completely subject to the influence of the external flow and we can thus focus on the effect of shape alone.

In the usual point-to-point navigation setup, the performance of a swimmer, self-propelling at constant speed v_0, is measured in terms of the total cost

$$\mathcal{C}_{\mathrm{p}} = \int_0^{t_{\mathrm{p}}} \left[\gamma_t v_0^2 + \beta\right] \mathrm{d}t = t_{\mathrm{p}} \left[\gamma_t v_0^2 + \beta\right] \tag{5.14}$$

which is analogous to the cost function (5.2) introduced above to quantify the smart throttling particle efficiency. However, here both the final time[6] t_{p} and the translational

[6] Indicated with t_{p} because it is the arrival time at the target for passive rotation, i.e. for self-propelled particles with a given shape whose orientation dynamics is entirely subject to the hydrodynamic torque (absence of active steering).

drag coefficient γ_t are fully determined by the shape parameter α. Moreover, here we also kept the free parameter β in order to conceptually distinguish between the two possible contributions to the total cost, namely energy dissipation and arrival time at the target.

In order to systematically study the impact of shape on the navigation efficiency, let us consider the two-dimensional shear flow setup already introduced in the previous section. The non-dimensionalized equations of motion now read

$$\begin{cases} \dot{x} = \zeta \cos\theta + y \\ \dot{y} = \zeta \sin\theta \\ \dot{\theta} = -\frac{1}{2}(1 - \alpha \cos 2\theta)\,, \end{cases} \tag{5.15}$$

with boundary conditions $\boldsymbol{r}(0) = \boldsymbol{0}$ and $\boldsymbol{r}(t_\mathrm{p}) = \hat{\boldsymbol{x}}$, and where the dimensionless parameter $\zeta \equiv v_0/(\kappa\ell)$ determines the system dynamics.

In Fig. 5.3a, we show three paths corresponding to microswimmers of different shapes in such setup. Already from this qualitative picture a key difference emerges: needle-like swimmers will tend travel along an almost straight line to the target (orange curve), while disk-like agents will make a greater use of the flow by moving further along the vertical direction, resulting in shorter travel times (green curve).

This qualitative insight can be easily verified by looking at the steady-state orientation θ_s of an active spheroid, namely imposing $\dot{\theta} = 0$ in Eq. (5.15). For $\alpha = -1$, we in fact find that such swimmer will tend to orient itself orthogonal to the flow ($\theta_\mathrm{s} = \pm\frac{\pi}{2}$), while a microswimmer with shape parameter $\alpha = 1$ will tend to align with it ($\theta_\mathrm{s} = 0, \pi$). For intermediate values of α, however, there is no stationary solution: in such cases, instead, the particles will perform a periodic tumbling motion, known as *Jeffery orbit* [5].

More quantitatively, we can then measure how the arrival times t_p at the target are affected by the shape of the swimmer by solving numerically[7] the corresponding ODE system (5.15) with the proper boundary conditions while varying α. Throughout our analysis, we will always compare swimmers with the same volume, which hereafter will be taken equal to $V = \frac{4\pi}{3}s^3$, with $s = 0.1\ell$.

The results obtained for three different values of ζ are reported in Fig. 5.3b. As analytically predicted, the arrival time is always minimal at $\alpha = -1$, and the relative improvement with respect to a needle-like swimmer ($\alpha \to 1$) becomes more and more significant as $\zeta \to 0$, going up to $\simeq 40\%$ for $\zeta = 0.1$. The differences among agents of different shapes are enhanced when $\zeta \ll 1$ because it corresponds to the regime where their motion is overall more affected by the presence of the external flow, for e.g. due to a small self-propulsion speed v_0 or strong shear rate κ.

However, as shown in Fig. 5.2, one observes the opposite trend when looking at the values of the translational drag coefficients γ_t for spheroids as function of α. Therefore, if we consider the overall navigation cost $\mathcal{C}_\mathrm{p}$ in Eq. (5.14), we have a clear trade-off coming from their shape. In order to illustrate this, we have therefore

[7] Exact solutions of this system are only available for $\alpha = [-1, 0, 1]$.

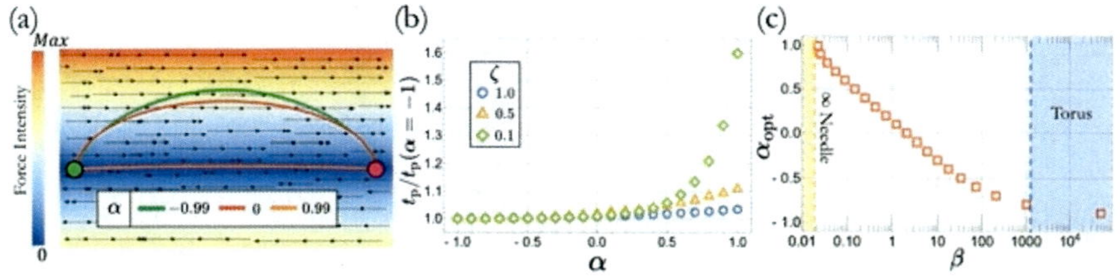

Fig. 5.3 Performance assessment of spheroidal swimmers in a shear flow. **a** Sample trajectories of three spheroids with different shapes: flat disk (green), sphere (red) and needle (orange). The latter points almost straight to the target (magenta circle) and does not make use of the flow, thus taking longer to reach it compared to the self-propelling disk. Here, the colour gradient represents the flow intensity with the black arrows indicating its direction. **b** Arrival times t_p at the target as function of the shape parameter α normalized with the time taken by a flat disk (corresponding to Zermelo's solution) for three different values of ζ. Note the relative increase in the arrival time for values of $\alpha > 0$, which gets more pronounced as $\zeta \to 0$. **c** Optimal value of the shape parameter α varying β, i.e. the relative weight between energy and time in the total cost C_p. Below a threshold the infinitely long needle is the most efficient shape ($\beta \lesssim 0.02$, shaded yellow area), while above $\beta \gtrsim 1400$ (shaded blue area) the optimal shape corresponds to a thin torus ($r/R = 0.1$). Data shown in **a** and **c** have been obtained at $\zeta = 0.1$

selected the optimal shape parameter α_{opt} minimizing the total cost C_p while varying the free parameter β representing the relative weight between energy consumption and travel time.

The corresponding results obtained at $\zeta = 0.1$ are shown in Fig. 5.3c. On the one hand, for small values of β, the particle is allowed to use little energy to navigate, such that the optimal shape is closer to a needle-like swimmer. On the other hand, as β is increased the optimal α shifts towards negative values[8], getting closer and closer to being a flat disk.[9]

However, there is a way to work around the strong inefficiency issue of a flat disk. We can indeed consider a toroidal object like the one shown in Fig. 2.5. Remarkably, a microswimmer shaped like a thin torus, i.e. when $r/R \ll 1$ in this representation, will undergo the same Jeffery rotation as a flat disk [17] (thus reaching the target in the shortest time) while featuring a significantly smaller translational drag. To compare quantitatively their performance, we have therefore computed the thin torus drag coefficient for active swimming in the axial direction by locally approximating it to a cylinder. The solution reads

$$\gamma_{t,\text{torus}} = 16\pi^2 R\,, \tag{5.16}$$

[8] Note that we do not expect the monotonic trends observed in Fig. 5.3b, c to hold in any dynamical regime. For some values of ζ, as a direct effect of Jeffery orbits there may in fact be swimmers with shape parameter $-1 < \alpha < 1$ that reach the target later than the needle.

[9] Here, β spans almost 5 decades is also due to the fact that the ratio between the translational drag coefficients of a flat disk and a needle is of the order $\sim 10^3$.

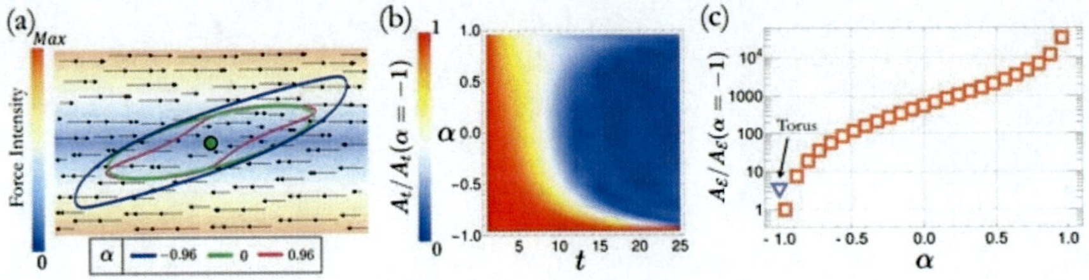

Fig. 5.4 Isochrones and isoenergies analysis for spheroids in a shear flow. **a** Exemplary isochrones in a shear flow at time $t = 7$ (in units of the shear rate κ) for spheroids of three different shapes: needle-like (magenta), spherical (green) and disk-like (blue). In this panel the color gradient shows for the intensity of the external flow while the black arrows indicate its direction. **b** Color map of the isochrones area as functions of both time and α. At each time, the area values have been divided by the area of the isochrone corresponding to $\alpha = -1$ (optimal by design). **c** Isoenergies area as function of the shape parameter α. All points are normalized with the area of the isoenergy corresponding to $\alpha = -0.96$. The triangle here is the isoenergy area for a thin torus with $r/R = 0.1$ and same volume as the spheroids. Here, the total amount of energy available to each swimmer is equal to the one dissipated by a spheroid with $\alpha = -0.96$ in a time $t = 0.1$. All data shown in **a–c** have been obtained setting $\zeta = 0.1$

where the particle mobility is set to unity and the accuracy of our approximation was numerically checked by means of a Boundary Element Method (BEM) [18]. We have then determined the corresponding energy dissipation considering a torus with the same volume as the spheroid and setting $r/R = 0.1$. As a result, for values of β large enough this alternative shape turns out to be the most efficient (see shaded blue area in Fig. 5.3c).

Conversely, below a certain threshold ($\beta \lesssim 0.02$ for the selected value of ζ) the infinitely long needle is the most efficient shape (see shaded yellow area in Fig. 5.3c). However, we can use some simple scaling arguments to show that there exists a small enough value of ζ for which $\alpha = 1$ is no longer optimal even for $\beta = 0$. On the one hand, since it follows a straight path to the target, the arrival time for the infinite needle scales as $t_\mathrm{p}(\alpha = 1) \sim \zeta^{-1}$. On the other hand, when $\zeta \to 0$ we can readily see from our exact solution at $\alpha = -1$ (see Eqs. (1.24-1.25)) that the arrival time scales as $t_\mathrm{p}(\alpha = -1) \sim \zeta^{-1/2}$, thus diverging more slowly than the one for a needle. From Eq. (5.14), this readily implies that the cost vanishes faster for $\alpha = -1$ than $\alpha = 1$ in the small ζ limit. Hence, there exists a dynamical regime at finite ζ in which the flat disk, or equivalently the thin torus, is energetically more efficient than any other spheroid.

Overall, we can also get rid of the specific target and achieve a similar insight into the navigation efficiency of spheroidal swimmers by looking at the isochrones. An example is shown in Fig. 5.4a for three different shapes: disk-like swimmers ($\alpha = -0.96$ here) are indeed the ones exploring faster the surrounding space, with the respective isochrone (blue curve) always enclosing the ones corresponding to swimmers of other shapes.

As introduced in Chap. 3, the isochrones area, hereafter indicated with A_t, provides a measure of the ability of a swimmer to efficiently explore the surroundings. We have therefore computed this quantity for all values of $-1 \leq \alpha \leq 1$ as function of time. The corresponding results are summarized in the color map shown in Fig. 5.4b, where the colour codes the isochrone area normalized with its value for $\alpha = -1$, which is optimal by design. Here, we can distinguish two regimes. At small times ($t \lesssim 8$) the area decreases monotonously with increasing α, such that the needle is the shape performing the worst. However, at longer times, some local minima appear for intermediate values of α. This is most likely due to the periodic tumbling experienced by the spheroids. Namely, this is essentially telling us that if we wait long enough, we will start see the effect of Jeffery's orbits [5].

Lastly, we can use the same idea and compute the area of the so-called isoenergies, i.e. curves essentially delimiting the potentially explored region with a given amount of energy. This quantity, which hereafter we will indicate with $A_{\mathcal{E}}$, indeed provides an alternative measure of the relative difference among microswimmers of different shapes in terms of energetic efficiency. Therefore, setting $\beta = 0$ in (5.14) we can fix the amount of energy the swimmers can dissipate during their motion and then compute the area they have covered. The corresponding results are shown in Fig. 5.4c. Not surprisingly, the more elongated particles are those that, given the same amount of energy available, are better at exploring the surroundings, covering an area several orders of magnitude larger than that of a disk. Even considering a thin torus, despite improving the energy efficiency of a disk, is not enough to compete with prolate spheroids (see the violet triangle in the plot). However, due to the non-linear effects of Jeffery's orbits already observed in the isochrones analysis, the increasing trend of this curve may not stay the same for any value of the energy, such that the flat disk and the thin torus may improve their efficiency with respect to the other shapes. Moreover, since $\sqrt{A_{\mathcal{E}}} \sim \ell \sim \zeta^{-1}$, the large energy (or equivalently long times) limit is equivalent to $\zeta \to 0$. As discussed above, in this regime the flat disk eventually becomes energetically more favourable, such that we can expect the corresponding isoenergy area to become the largest.[10]

All in all, our study of navigation efficiency of a swimmer reveals a natural trade-off between energy consumption and time that emerges as a straightforward effect of accounting for the microswimmer's shape.

In this setup the active agents do not have any control over their dynamics and have been therefore referred to as *dumb* swimmers. However, in this way the overall efficiency is fully determined a priori by the flow configuration and the swimmer shape, such that there is no room for further optimization. To this end, we shall now take our analysis a step further and see how we can improve the performance of a swimmer with a well-defined shape by making it *smart*, namely by allowing it to actively steer during the navigation.

---

[10] However, this effect would be observed only for very long simulation times.

5.3 Navigation by Smart Swimmers: The (Energetic) Cost of Steering

Let us now consider an overdamped self-propelled spheroidal particle moving at constant speed v_0 in the presence of an external stationary flow $\boldsymbol{f}(\boldsymbol{r})$. Such swimmer is capable of controlling its own orientation dynamics by actively steering while being also subject to the flow-induced torques. As a result, the equations governing the microswimmer's motion will now read

$$\begin{cases} \dot{\boldsymbol{r}} = v_0 \hat{\boldsymbol{u}} + \boldsymbol{f}(\boldsymbol{r}) \\ \dot{\hat{\boldsymbol{u}}} = \boldsymbol{\omega}_a + \boldsymbol{\omega}_f(\boldsymbol{r}, \hat{\boldsymbol{u}}) \,, \end{cases} \tag{5.17}$$

with $\boldsymbol{\omega}_f$ being Jeffery's angular velocity (5.11) while $\boldsymbol{\omega}_a$ is the active angular velocity the swimmer can self-generate. Thanks to the introduction of this new degree of freedom in the swimmer dynamics, we may now use the tools from OC theory to study the possible performance improvement of the navigation for finite-size swimmers with a well-defined shape.

In this scenario, the task is to reach a specific target while minimizing the arrival time as well as the total energetic dissipation due to both the translational drag and the active steering. Namely, the overall navigation cost becomes

$$\mathcal{C}_s = \int_0^{t_f} \left(\gamma_r \omega_a^2 + \gamma_t v_0^2 + \beta\right) \mathrm{d}t \,. \tag{5.18}$$

where γ_r is the rotational drag coefficient for a surface-driven spheroidal swimmer, already shown in Fig. 5.2, which was obtained by combining the no-slip and perfect-slip coefficients [9, 11] via the same formula (5.13) used to determine the respective translational drag coefficient γ_t. The system Hamiltonian here thus takes the form

$$\mathcal{H} = \beta + \gamma_t v_0^2 + \gamma_r \omega_a^2 + \boldsymbol{p} \cdot \dot{\boldsymbol{r}} + \boldsymbol{p}_u \cdot \dot{\hat{\boldsymbol{u}}} + \mu \hat{\boldsymbol{u}} \cdot \hat{\boldsymbol{u}} \,, \tag{5.19}$$

where $\boldsymbol{p}_u$ are the momenta associated with the orientation $\hat{\boldsymbol{u}}$, which in this context plays the role of a state variable together with the position vector $\boldsymbol{r}$. To ensure the vector $\hat{\boldsymbol{u}}$ has constant norm during the motion, we have also added a constraint in the Hamiltonian which corresponds to the last term in (5.19), with μ being a Lagrange multiplier.

In the OC framework, we can identify the active angular velocity $\boldsymbol{\omega}_a$ as the control variable and, based on Pontryagin's minimum principle (1.10), we may now write all the necessary conditions for optimality

$$\begin{cases} \dot{p}_i = -p_j \partial_i f_j - (p_{u,j} + \mu \hat{u}_j) \partial_i \omega_{f,j} \\ \dot{p}_{u,i} = -v_0 p_i - (p_{u,j} + \mu \hat{u}_j) \partial_{u_i} \omega_{f,j} - \mu \omega_{a,i} \\ \frac{\partial \mathcal{H}}{\partial \omega_{a,i}} = 0 \Longrightarrow \omega_{a,i} = -\frac{p_{u,i} + \mu \hat{u}_i}{2\gamma_r} \\ \mathcal{H} = 0 \,, \end{cases} \tag{5.20}$$

where $\partial_i \equiv \partial_{r_i}$, and with the last condition only necessary if we are not specifying the final time t_f. In the presence of a specific target $\boldsymbol{r}_T$ and without constraining the initial and final orientation, the boundary conditions to equip this system with are:

$$\begin{cases} \boldsymbol{r}(0) = \boldsymbol{r}_0\,, \quad \boldsymbol{r}(t_f) = \boldsymbol{r}_T \\ \boldsymbol{p}_u(0) = \boldsymbol{p}_u(t_f) = 0\,. \end{cases} \tag{5.21}$$

Note that the boundary conditions on the momenta are there if and only if the corresponding state variable is unconstrained at the initial and/or final times. Finally, in order to satisfy the constraint on the orientation vector $\hat{\boldsymbol{u}}$, from the third equation in (5.20) we can see that the Lagrange multiplier μ must be chosen as

$$\mu = -\boldsymbol{p}_u \cdot \hat{\boldsymbol{u}}\,. \tag{5.22}$$

5.3.1 Smart Steering in a Shear Flow

We shall now illustrate the implementation of the smart steering protocol for spheroidal swimmers in the 2D shear flow setup shown in Fig. 5.3a. As a result, the system of equations to be solved simplifies[11]:

$$\begin{cases} \dot{x} = \zeta \cos\theta + y \\ \dot{y} = \zeta \sin\theta \\ \dot{\theta} = \omega_a - \frac{1}{2}(1 - \alpha\cos 2\theta) \\ \dot{p}_x = 0 \\ \dot{p}_y = -p_x \\ \dot{\omega}_a = \frac{\zeta}{\Gamma}\, \boldsymbol{p} \cdot \hat{\boldsymbol{u}}^{\perp} + \omega_a \alpha \sin 2\theta\,, \end{cases} \tag{5.23}$$

where we have rescaled space, time and the momenta respectively as $\boldsymbol{r} \to \ell \boldsymbol{r}$ and $t \to t/\kappa$ and $\boldsymbol{p} \to 2\gamma_t \kappa \ell \boldsymbol{p}$. The system dynamics is therefore determined by the two dimensionless parameters $\zeta \equiv v_0/(\kappa\ell)$ and $\Gamma \equiv \gamma_r/(\gamma_t \ell^2)$. Note that the latter is a quantity essentially determined by the swimmer shape and size, while we still have freedom in choosing a value for the former.

As already shown in Fig. 5.3b, varying ζ amounts to exploring two possible regimes: a flow dominated one ($\zeta < 1$), where the differences in terms of performance between swimmers of different shape are enhanced, and the opposite one dominated by self-propulsion ($\zeta > 1$), where these effects become negligible. The first regime is thus the most interesting to explore in terms of optimal navigation.

[11] In 2D there is no need to use the Lagrange multiplier μ as we can directly deal with the orientation angle θ rather than with the heading direction vector $\hat{\boldsymbol{u}}$. Moreover, this simplification allows to directly obtain an equation for the control variable ω_a.

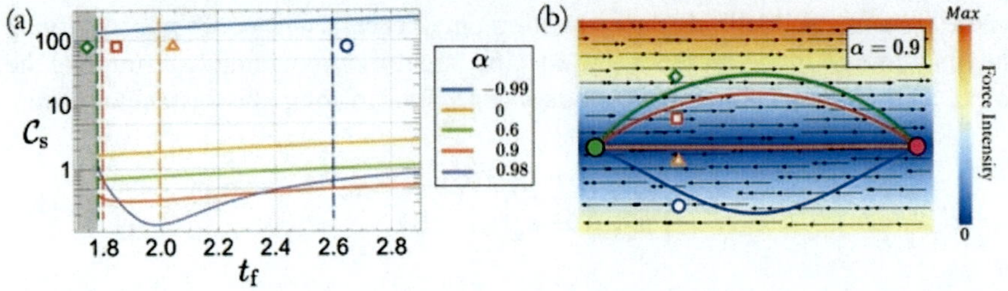

Fig. 5.5 Swimming cost of smart active spheroids in a shear flow. **a** Minimal overall energetic cost $\mathcal{C}_s$ of navigation for smart swimmers of different shapes as function of the arrival time t_f at the target. All curves stop at the optimal time $t_{opt} = 1.7854$ (in units of the inverse shear rate κ^{-1}), below which it is not possible to reach the target (shaded gray area). **b** Optimal trajectories for a spheroid with shape parameter $\alpha = 0.9$ corresponding to different arrival times as indicated by the corresponding symbols in **a**. All the paths start from the origin (green circle) and end at the target in position $\boldsymbol{r}_T = \hat{\boldsymbol{x}}$ (magenta circle). The color gradient here shows the flow intensity, while the black arrows its direction. All data shown in **a**, **b** have been obtained setting $\zeta = 0.5$

Therefore, in the following we set[12] $\zeta = 0.5$ by effectively halving the swimmer self-propulsion speed v_0 while keeping both the shear rate κ and the target distance ℓ equal to unity.

Our goal is now to determine for each value of the shape parameter α the optimal steering protocol $\omega_a(t)$ that minimizes the total cost (5.18), which can also be rewritten in units of $\gamma_t \kappa \ell^2$ as

$$\mathcal{C}_s = \Gamma \Omega_a(t_f) + \left(\zeta^2 + \rho\right) t_f \,, \tag{5.24}$$

where we have concisely defined $\Omega_a(t_f) \equiv \int_0^{t_f} \omega_a^2 \mathrm{d}t$ and $\rho \equiv \frac{\beta}{\gamma_t (\kappa \ell)^2}$.

However, in contrast with the smart throttling case illustrated in Sect. 5.1.1 of a point-like particle moving in the same flow configuration, the dynamical system (5.23) is not solvable analytically. Therefore, since we do not know a priori what the optimal arrival time t_f at the target is for each value of the shape parameter α, we shall solve the system numerically by scanning over values of t_f, reaching the minimum one allowed, i.e. the arrival time t_{opt} obtained from Zermelo's solution.[13] Furthermore, as this is a boundary value problem, we must resort to refined numerical shooting methods in order to find the optimal solution for each combination of α and t_f (details in A.3). Note that, in the following, we compare swimmers of different shapes while keeping their volume constant and equal to $V = \frac{4\pi}{3} s^3$, with $s = 0.1\ell$.

The corresponding results obtained from our numerical simulations are shown in Fig. 5.5a. There, we show the energetic cost of navigation (in dimensional units), i.e. $\mathcal{C}_s$ defined in Eq. (5.18) with $\beta = 0$, as function of the arrival time t_f. Thanks to

---

[12] A value small enough such that we do not expect our results to qualitatively change decreasing further ζ.

[13] Although in (5.24) we are still expressing the total cost including the explicit contribution from the arrival time at the target, in the following analysis we shall set $\beta = 0$ as it would otherwise just amount to shift all the cost curves by the same quantity, thus not qualitatively changing our results.

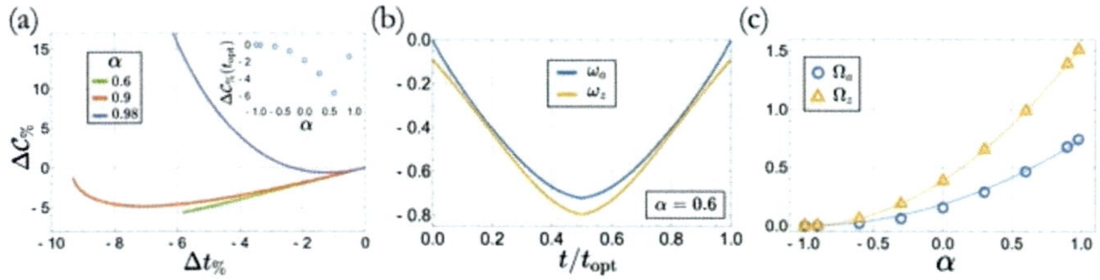

Fig. 5.6 Active steering efficiency of spheroidal swimmers in a shear flow. **a** Relative variation in percentage of the energetic cost $\Delta C_{\%}$ of navigation as function of the percentage of time $\Delta t_{\%}$ saved to reach the target with respect to the *dumb* swimming case, i.e. without active steering, for three exemplary values of α. Inset: Percentage variation in the energetic cost between *dumb* swimming and *smart* navigation, i.e. when the microswimmer instead actively steers to reach the target in the shortest time $t_{\rm opt}$ as function of the shape parameter α. **b** Active angular velocity ω_a implemented by the smart swimmer with shape parameter $\alpha \approx 0.6$ to get as fast as possible to the target while saving energy as function of time (blue curve). Here, it is compared with the steering protocol ω_z (orange curve) to be implemented so as to make the swimmer exactly follow Zermelo's path. **c** Comparison between the non-dimensionalized costs of steering associated to the optimal active control ω_a (blue circles) and Zermelo's protocol ω_z (orange triangles) as function of α. Both of them make the swimmer reach the target in the shortest time $t_{\rm opt}$, but at a different overall cost. Solid lines here represents the corresponding quadratic fits. All data shown in **a–c** have been obtained setting $\zeta = 0.5$

the observed non-monotonic trends, there exist some crossovers between the curves both at long and short times, such that we can identify different optimal shapes for a spheroidal swimmer depending on the desired arrival time $t_{\rm f}$.

Some exemplary optimal trajectories for a spheroid with $\alpha = 0.9$ are shown in Fig. 5.5b. When the task is to reach the target in a time $t_{\rm f} < t_{\rm p}$ the swimmer will make more and more use of the flow, thus moving further in the vertical direction (green and red curves). Conversely, when the arrival time is such that $t_{\rm f} > t_{\rm p}$, the energetically most convenient thing to do is to point more straight (orange curve) or even go in the counter-flow region if necessary (typically at longer times, see blue curve instead).

Thanks to their active rotation, swimmers of any shape can thus now reach the target faster than in the *dumb* swimming case studied in Sect. 5.2.2. However, this comes at the cost of spending some extra energy to steer. Hence, this raises the question of which the relative variation in the overall cost actually is for a microswimmer trying to improve its time performance with respect to the *dumb* swimming case. This can be quantified as

$$\Delta C_{\%} \equiv \frac{(C_{\rm s} - C_{\rm p})}{C_{\rm p}} \cdot 100 = \Delta t_{\%} + \frac{\Gamma}{t_{\rm p}} \Omega_a(t_{\rm f}) \cdot 100 \,, \tag{5.25}$$

which is the percentage cost variation with respect to the *dumb* swimming case (setting $\beta = 0$), and where we have also defined the percentage improvement of time as $\Delta t_{\%} \equiv \frac{(t_{\rm f} - t_{\rm p})}{t_{\rm p}} \cdot 100$.

The plot in Fig. 5.6a shows the corresponding results for three different values of the shape parameter α. On the one hand, we find that, for small improvements in time, swimmers of any shape manage to save a proportional amount of energy. On the other hand, even more remarkably, it turns out that some shapes feature a relative gain curve with a monotonic decreasing trend, which means that saving time always translates into spending overall less energy. This is the case, e.g., for a swimmer with $\alpha = 0.6$ (green curve), which improves its time performance by $\Delta t_{\%} \simeq -6\%$ (reaching the target at $t = t_{\text{opt}}$), and moreover manages to save relatively circa the same amount of energy. Conversely, swimmers with more elongated shapes like $\alpha = 0.98$ (violet curve), need to spend comparatively much more energy than in the *dumb* swimming case ($\Delta C_{\%}$ can reach $\simeq +200\%$) in order to reach the target in the shortest time t_{opt}.

This counter-intuitive effect is possible because, even though reducing the navigation time t_{f} requires the microswimmer to actively steer and thus spend more rotational energy, by doing so it also lowers the translational drag dissipation which is indeed proportional to t_{f}. We are thus left with an emerging trade-off purely based on the relative weight between the rotational and translational energy costs, which ultimately depends just on the swimmer shape and size.

Therefore, it is not surprising that we find an optimal value of the shape parameter α that maximises the energy savings when willing to reach the target in the shortest time t_{opt}. As shown in the inset in Fig. 5.6a, the corresponding optimum is here achieved at $\alpha \approx 0.6$.

Finally, we can also look at these results from a different perspective. In fact, one could have naively guessed that the optimal steering strategy that allows the swimmer to reach the target in the shortest time would essentially amount to

$$\omega_z(\alpha) = -\omega_f(\alpha) + \omega_f(-1)\,, \tag{5.26}$$

which is essentially the generalization of Zermelo's protocol (second equation in system (5.5)) to particles with a well-defined shape. Here, ω_f is indeed the usual Jeffery angular velocity (5.11) which we know is equal to Zermelo's solution for $\alpha = -1$, such that by implementing the policy (5.26) a microswimmer of any shape would exactly follow Zermelo's path, thus reaching the target at time t_{opt} by definition. However, the smart steering strategy ω_a obtained from OC theory does not converge to ω_z when $t_{\text{f}} \to t_{\text{opt}}$[14] for values of the shape parameter $\alpha \neq -1$. As shown in Fig. 5.6b for an exemplary shape, these two policies are indeed intrinsically different: ω_a is by design equal to zero at the initial and final times, while ω_z cannot be unless $\alpha = -1$.

We can thus compare the two steering costs associated with each of these two strategies. Since they both share the same prefactors, it is sufficient to consider just $\Omega_a(t_{\text{opt}}) \equiv \Omega_a = \int_0^{t_{\text{opt}}} \omega_a^2 dt$ and $\Omega_z \equiv \int_0^{t_{\text{opt}}} \omega_z^2 dt$. In the shear flow setup, the latter can be actually computed analytically. Starting from the ODE system (5.23) and using Eq. (1.24), we get

[14] In our simulations, for each α we have actually managed to scan arrival times down to values just 0.03% larger than t_{opt}.

$$\Omega_z = \frac{(1+\alpha)^2}{2}\left[\tan^3\theta_0 + 3\tan\theta_0 - 2\theta_0(1+\tan^2\theta_0)\right], \tag{5.27}$$

which together with (1.25) makes up the exact solution, where θ_0 is the proper initial angle the swimmer should take so as to get to the target in the shortest time.

The corresponding curves as function of the shape parameter α are shown in Fig. 5.6c. They both show a quadratic trend (solid curves stand for the respective fits) and, remarkably, the smart active steering ω_a always leads to less energy consumption, improving the performance of Zermelo's strategy ω_z up to a factor 2.

Hence, thanks to our analysis based on the results from OC theory, we have here effectively found a way to outperform Zermelo's solution. In fact, not only can these smart swimmers reach the target in the theoretically shortest time, but they also manage to reduce the corresponding energetic cost when compared to Zermelo's policy.

5.4 Summary and Discussion

We have here shown some possible extensions to the problem of optimal navigation of microswimmers.

Firstly, we have considered swimmers which can tune their own self-propulsion speed. Thus, using tools from OC theory we have derived the smart throttling protocol which minimizes the weighted sum of translational drag dissipation and arrival time at the target. Moreover, we have been able to provide an exact solution to this problem when navigation takes place in the presence of a linear shear flow. Then, depending on the relative weights between energy consumption and time in the overall cost, we have identified two different regimes in which the swimmer will preferably tend to minimize one or the other quantity, leading to more curved or straighter trajectories, respectively.

We have then taken into account spheroidal microswimmers with a finite size and therefore considered for the first time how the external flow affects the swimmers orientation dynamics. Since in this context they do not have any control over their dynamics, we have called them *dumb* swimmers.

Remarkably, the orientation dynamics corresponding to Zermelo's solution, i.e. the one minimizing travel time by design, turns out to be the same as the one of a self-propelling flat disk in the presence of a stationary flow. On the one hand, an active swimmer with such shape will thus always automatically follow the time-optimal path without the need for active steering. On the other hand, however, a needle-like swimmer typically dissipates much less energy during its motion, such that we have here revealed a natural trade-off which depends only on the microswimmer shape. This is not only relevant for point-to-point navigation, but also for more generic optimal exploration problems, as it has been confirmed by our analysis of the isochrones and isoenergies areas in the shear flow setup.

As a result, depending on the relative weight between navigation time and energy consumption in the total cost function, we can always identify an optimal shape for the swimmer that maximizes the overall swimming efficiency. Furthermore, we have shown that the difference in the arrival times between a flat disk and a long needle increases with the relative flow strength, i.e. smaller self-propulsion speed or larger shear rate, such that there exists a strong flow regime for which more elongated swimmers would actually be less energetically efficient than a flat disk.

However, in order to further optimize its own navigation, a microswimmer of any shape needs to have some control over its motion. To this end, we have therefore generalized the problem of optimal navigation to particles with a well-defined shape using OC theory. In this way, we could then once again determine the set of equations to be solved in order to find the optimal steering strategy minimizing the overall cost of navigation, which by then included also a cost associated with the active re-orientation.

Consequently, we have found out that depending on the desired arrival time at the target, there is an optimal shape for a swimmer to be considered. Moreover, thanks to the new degree of freedom, swimmers of any shape can reach the target in the shortest time. Even more remarkably, by improving their time performance with respect to the *dumb* swimming case, these microswimmers also automatically reduce their energy consumption, such that being smart turns out to have a double advantage.

Lastly, our analysis has revealed that the smart steering strategies obtained from OC theory do not naively boil down to Zermelo's policy when the swimmers want to reach the target in the shortest time. In fact, we have shown that swimmers implementing such steering protocols manage to save more energy than in the latter case. In other words, by making the swimmers smart, we have allowed them to perform better than by being forced to follow Zermelo's strategy, which is not explicitly based on energy minimization principles.

References

1. Liebchen B, Löwen H (2019) Optimal navigation strategies for active particles. EPL (Eur Lett) 127(3):34003. https://doi.org/10.1209/0295-5075/127/34003
2. Pinti J, Celani A, Thygesen UH, Mariani P (2020) Optimal navigation and behavioural traits in oceanic migrations. Theor Ecol 13(4):583–593. https://doi.org/10.1007/s12080-020-00469-4
3. Bretherton FP (1962) The motion of rigid particles in a shear flow at low reynolds number. J Fluid Mech 14(2):284–304. https://doi.org/10.1017/S002211206200124X
4. Kaya T, Koser H (2009) Characterization of hydrodynamic surface interactions of escherichia coli cell bodies in shear flow. Phys Rev Lett 103:138103
5. Jeffery GB, Filon LNG (1922) The motion of ellipsoidal particles immersed in a viscous fluid. Proc R Soc Lond Ser A Contain Pap Math Phys Character 102(715):161–179. https://royalsocietypublishing.org/doi/abs/10.1098/rspa.1922.0078
6. Arguedas-Leiva J-A, Wilczek M (2020) Microswimmers in an axisymmetric vortex flow. New J Phys 22(5):053051. https://doi.org/10.1088/1367-2630/ab776f

7. Chen X-Z, Jang B, Ahmed D, Hu C, De Marco C, Hoop M, Mushtaq F, Nelson BJ, Pané S (2018) Small-scale machines driven by external power sources. Adv Mater 30(15):1705061. https://doi.org/10.1002/adma.201705061
8. Mano T, Delfau J-B, Iwasawa J, Sano M (2017) Optimal run-and-tumble based transportation of a janus particle with active steering. Proc Natl Acad Sci 114(13):E2580–E2589. https://doi.org/10.1073/pnas.1616013114
9. Chang YC, Keh HJ (2009) Translation and rotation of slightly deformed colloidal spheres experiencing slip. J Colloid Interface Sci 330(1):201–210. ISSN 0021-9797. https://doi.org/10.1016/j.jcis.2008.10.055, https://www.sciencedirect.com/science/article/pii/S0021979708013039
10. Happel J, Brenner H (1983) Low Reynolds number hydrodynamics. Springer, Netherlands, Dordrecht
11. Hu C-M, Zwanzig R (1974) Rotational friction coefficients for spheroids with the slipping boundary condition. J Chem Phys 60(11):4354–4357. https://doi.org/10.1063/1.1680910
12. Blake J (1973) A finite model for ciliated micro-organisms. J Biomech 6(2):133–140. ISSN 0021-9290. https://doi.org/10.1016/0021-9290(73)90082-1, https://www.sciencedirect.com/science/article/pii/0021929073900821
13. Blake JR (1971) A spherical envelope approach to ciliary propulsion. J Fluid Mech 46(1):199–208. https://doi.org/10.1017/S002211207100048X
14. Golestanian R, Liverpool TB, Ajdari A (2005) Propulsion of a molecular machine by asymmetric distribution of reaction products. Phys Rev Lett 94:220801. https://doi.org/10.1103/PhysRevLett.94.220801
15. Golestanian R (2022) Phoretic active matter. In: Active matter and nonequilibrium statistical physics, Lecture notes of the (2018) Les Houches summer school. Oxford University Press, London, England
16. Nasouri B, Vilfan A, Golestanian R (Jan2021) Minimum dissipation theorem for microswimmers. Phys Rev Lett 126:034503
17. Singh V, Koch DL, Stroock AD (2013) Rigid ring-shaped particles that align in simple shear flow. J Fluid Mech 722:121–158. https://doi.org/10.1017/jfm.2013.53
18. Hall WS (1994) Boundary element method. Springer, Netherlands, Dordrecht, pp 61–83. ISBN 978-94-011-0784-6. https://doi.org/10.1007/978-94-011-0784-6_3

Chapter 6
Concluding Remarks

I don't know anything, but I do know that everything is interesting if you go into it deeply enough.

Richard P. Feynman

Given the great hope it gives to tackle important challenges in our society across various contexts, the problem of optimal navigation has recently attracted increasing interest among scientists from different communities. The study of the optimization of a self-propelling agent's motion in a complex environment is, moreover, relevant not only for the obvious practical applications, but also from an ethological perspective. The predictive potential of theories and models coming from all branches of physics may indeed help to decipher the complex dynamics behind living systems behaviors [1, 2].

Although considerable progress has been made over the years since Zermelo's seminal work in 1931, there are still a number of pending issues in this field. Here, we have addressed some of them by providing new theoretical and numerical tools that allow for a more complete study of the optimal navigation problem for active particles.

In Chap. 2, we have analytically solved the problem of optimal navigation on curved surfaces and in the presence of stationary flows. By formulating a geometric description of the problem, we have shown that the minimal-time path are essentially geodesics in a Finsler space described by a Randers metric. Remarkably, this formalism has the additional advantage of being coordinate-free and also compact enough so as to greatly simplify the analysis of this problem and its potential generalizations.

A first extension of the model introduced here would certainly aim to include Brownian noise into its description, possibly by using the Onsager-Machlup functional [3, 4]. Furthermore, this new bridge between microswimmer physics and general relativity could enrich both fields in the future. For example, one can consider finite-size swimmers able to sense curvature by estimating spatial derivatives of the metric [5]. This would in turn open the door to the study of the corresponding geodesic

L. Piro, *Optimal Navigation in Active Matter*, Springer Theses,
https://doi.org/10.1007/978-3-031-52577-3_6

deviation equation [6], thus potentially providing new insights into the stability and irreversible character of an optimal path depending on the geometrical properties of the space. Further research might also use this analytical approach to benchmark numerical optimization procedures relying on machine learning algorithms [7, 8].

In Chap. 3, we then studied the efficiency of both Zermelo's classical solution in flat space and Randers geodesics on curved surfaces, focusing on the regime where the external force exceeds self-propulsion in finite regions of a potential landscape. Thanks to the extensive use of isochrones as new powerful tool for analysis, we were there able to characterize in more detail such navigation strategies. On the one hand, our results show that they indeed allow the active particle to comprehensively explore the environment, as opposed to the more trivial strategy of going straight, whose most trajectories get arrested in some local minima of the potential. On the other hand, our analysis also indicates an increased sensitivity of these time-minimizing strategies to initial conditions, a feature which limits their robustness and long-time efficiency, especially in presence of fluctuations. Hence, these results suggest an interesting trade-off between exploration efficiency and stability for the design of control strategies to be implemented in real systems.

Since the chaoticity of Zermelo's and Randers' policies turns out to be closely linked to the presence of isochrone self-crossings occurring in strong force regions, inhibiting these seems crucial in order to improve the strategies stability and robustness. A possibility could thus be to include an additional cost in the derivation of such protocols which would lead the agents to reorient their self-propulsion when entering strong force regions. A further improvement could also be achieved allowing for some degree of variation of the particles self-propulsion speed. Such feature indeed would provide the particles with a mechanism to visit otherwise inaccessible regions while inhibiting self-crossings at minimal energetic cost.

Despite studying the effect of strong force fields on the active particle navigation, here we did not discuss the trapping regime where the particles must rely on fluctuations to cross a potential barrier. It has been argued that in the presence of confinement induced e.g. by obstacles [9, 10] or high potential barriers [11, 12], efficient exploration strategies favor diffusive over ballistic motion. However, in all these works no optimization policy making use of the environmental conditions was considered, which makes this still a promising direction for future research.

In Chap. 4, we have instead addressed the problem of preserving the active agent's autonomy when it moves in a complex environment and in the presence of noise. Starting from the solution provided by stochastic optimal control (SOC) theory, we were able to identify some simple general principles on which it relies. Specifically, it turned out that the optimal navigation strategy in the presence of fluctuations essentially amounts to stabilizing the motion of the active particle around the noiseless optimal path, i.e. the one provided by Zermelo's solution, while maximizing the drift along it. This observation therefore inspired us to design a new set of navigation strategies that allows the micro-swimmer to navigate semi-autonomously and, at the same time, efficiently towards a target. While relying on much less information and simple alignment rules with local gradients and flow, these newly-designed strategies show performances surprisingly similar to those achieved by the optimal policy from SOC theory. Furthermore, we have shown their robustness to the presence of both strong translational and rotational fluctuations, and also their great versatility by implementing them on curved surfaces and in random flows.

Although the analysis carried out here focused on the problem of travel time optimization, the policies we proposed are fully determined by the noiseless optimal path. Therefore, they are straightforwardly generalizable to a broader class of optimization problems like energy dissipation or fuel consumption minimization. Moreover, it would be very interesting to implement reinforcement learning-based algorithms in this context, so as to check whether they would reach similar conclusions regarding the core principles underlying an efficient navigation in the presence of fluctuations. This would in fact further confirm the validity and generality of our arguments.

In Chap. 5, we have finally further extended Zermelo's classical problem by relaxing some assumptions of the model, so as to get closer to real-life systems. To this end, we firstly allowed the swimmer to control its own self-propulsion speed, which in turn introduced a trade-off between arrival time at the target and energy consumption. We have then investigated how navigation in a fluid flow is affected when considering microswimmers with a well-defined shape. Studying the exemplary case of active spheroids in a shear flow, we have interestingly revealed an emerging trade-off between time and energy dissipation solely depending on the swimmers shape. Our results indeed show that swimmers without any control over their motion, hence referred to as *dumb*, can reach a target faster if their shape is more flattened, i.e. like a disk, but at a higher energetic cost when compared to needle-like swimmers. By allowing them to actively steer during their motion, we have then made these agents *smart*. Remarkably, our analysis reveals that such swimmers are not only capable of reaching the target faster with respect to the passive case, but also in doing so they spend overall less energy. Furthermore, as the arrival time approaches Zermelo's solution, the smart steering strategies do not converge to Zermelo's policy as naively expected. In fact, it turns out they outperform it: although reaching the target at the same (optimal) time, such smart swimmers manage to dissipate systematically less energy, regardless of their shape.

Our aim in the future is to consider even *smarter* swimmers able to both control their self-propulsion speed and actively steer, so as to possibly uncover basic navigation principles which would further help in designing smart artificial swimmers. Moreover, it would be interesting to take into account swimmers with different shapes other than spheroids, which may enhance their navigation efficiency even further [13, 14], and also test the robustness of our results by considering more complex flow configurations.

As a general outlook, although here we have focused on the situation in which navigation is performed by a single agent, it would be definitely worth studying the optimization of collective processes by including interactions among multiple agents [15, 16]. In fact, cooperation between agents has already been proved useful to improve the overall performance [17, 18], even in the presence of just a few leaders [19, 20] who efficiently manage to guide the group towards the achievement of common tasks.

The study of collective optimal navigation in active matter additionally raises the challenging issue of the possible presence of hydrodynamic interactions between the agents [21]. To the best of our knowledge, this is a field of research yet to be explored as to date there is only one example, see Ref. [22], where these have been considered in the two-dimensional case of a microswimmer interacting with simple obstacles.

Another possible future direction would be the study of non-Markovian navigation strategies. The use of memory by an agent certainly constitutes an additional degree of complexity, yet could allow it to perform increasingly complex tasks [23] and even alter its navigation strategy depending on the amount of information it is capable of accumulating over time [24].

Overall, it would also be useful to further extend the robustness study regarding the navigation strategies discussed here by including a time delay in the active particle orientation [25]. This would indeed help to mimic what happens in actual experiments in which the protocol actuation takes place via external feedback control [26, 27].

Moreover, the analysis of the efficiency of the navigation policies can be further refined. It would indeed be interesting to study what are the thermodynamic costs related to the assimilation of information [28, 29] by an active agent during its motion. Since higher complexity translates into higher thermodynamic cost[1], the strategies that require more information from the environment—and typically perform better—would then be penalised. Taking this new cost into account could therefore significantly reshape our general considerations on the efficiency of a navigation strategy.

All in all, the advances made in this thesis have both significantly extended the state of the art in optimal navigation for active matter systems, but also opened up a number of new possible directions to be pursued in future research.

References

1. Pezzotta A, Adorisio M, Celani A (2018) Chemotaxis emerges as the optimal solution to cooperative search games. Phys Rev E 98:042401. https://doi.org/10.1103/PhysRevE.98.042401
2. Pinti J, Celani A, Thygesen UH, Mariani P (2020) Optimal navigation and behavioural traits in oceanic migrations. Theor Ecol 13(4):583–593. https://doi.org/10.1007/s12080-020-00469-4
3. Onsager L, Machlup S (1953) Fluctuations and irreversible processes. Phys Rev 91:1505–1512. https://doi.org/10.1103/PhysRev.91.1505
4. Woillez E, Zhao Y, Kafri Y, Lecomte V, Tailleur J (2019) Activated escape of a self-propelled particle from a metastable state. Phys Rev Lett 122:258001. https://doi.org/10.1103/PhysRevLett.122.258001
5. Bao D, Robles C (2004) Ricci and flag curvatures in finsler geometry. A Sampler of Riemann-Finsler Geometry 50:01
6. Stavrinos PC (1993) Generalized-finslerian equation of geodesic deviations. Rep Math Phys, 32(3):339–342. ISSN 0034-4877. https://doi.org/10.1016/0034-4877(93)90025-A, http://www.sciencedirect.com/science/article/pii/003448779390025A

[1] The degree of sophistication of a strategy can indeed be formally measured in terms of thermodynamic quantities [30], such as work and entropy.

7. Muiños-Landin S, Fischer A, Holubec V, Cichos F (2021) Reinforcement learning with artificial microswimmers. Sci Robot 6(52):eabd9285. https://doi.org/10.1126/scirobotics.abd9285
8. Cichos K, nnd Gustavsson F, Mehlig B, Volpe G, (2020) Machine learning for active matter. Nat Mach Intell 2:94–103. https://doi.org/10.1038/s42256-020-0146-9
9. Volpe G, Volpe G (2017) The topography of the environment alters the optimal search strategy for active particles. Proc Natl Acad Sci 114(43):11350–11355. https://doi.org/10.1073/pnas.1711371114
10. Irani E, Mokhtari Z, Zippelius A (2022) Dynamics of bacteria scanning a porous environment. Phys Rev Lett 128:144501. https://doi.org/10.1103/PhysRevLett.128.144501
11. Zanovello L, Caraglio M, Franosch T, Faccioli P (2021) Target search of active agents crossing high energy barriers. Phys Rev Lett 126:018001. https://doi.org/10.1103/PhysRevLett.126.018001
12. Zanovello L, Faccioli P, Franosch T, Caraglio M (2021) Optimal navigation strategy of active brownian particles in target-search problems. J Chem Phys 155(8):084901. https://doi.org/10.1063/5.0064007
13. Vilfan A (2012) Optimal shapes of surface slip driven self-propelled microswimmers. Phys Rev Lett 109:128105. https://doi.org/10.1103/PhysRevLett.109.128105
14. Guo H, Zhu H, Liu R, Bonnet M, Veerapaneni S (2021) Optimal slip velocities of micro-swimmers with arbitrary axisymmetric shapes. J Fluid Mech 910:A26. https://doi.org/10.1017/jfm.2020.969
15. Borra F, Cencini M (2021) Celani A (2021) Optimal collision avoidance in swarms of active brownian particles. J Stat Mech: Theory Exp 8:083401. https://doi.org/10.1088/1742-5468/ac12c6
16. Berekméri E, Zafeiris A (2020) Optimal collective decision making: consensus, accuracy and the effects of limited access to information. Sci Rep. https://www.nature.com/articles/s41598-020-73853-z
17. Bonabeau E, Theraulaz G, Dorigo M (1999) Swarm intelligence: from natural to artificial systems. Oxford University Press, New York, USA, p 0195131592
18. Durve M, Piro L, Cencini M, Biferale L, Celani A (2020) Collective olfactory search in a turbulent environment. Phys Rev E 102:012402. https://doi.org/10.1103/PhysRevE.102.012402
19. Garland J, Berdahl AM, Sun J, Bollt EM (2018) Anatomy of leadership in collective behaviour. Chaos: Interdiscip J Nonlinear Sci 28(7):075308. https://doi.org/10.1063/1.5024395
20. Gómez-Nava L, Bon R, Peruani F (2022) Intermittent collective motion in sheep results from alternating the role of leader and follower. Nat Phys. https://www.nature.com/articles/s41567-022-01769-8
21. Gompper G, Bechinger C, Herminghaus S, Isele-Holder R, Kaupp UB, Löwen H, Stark H, Winkler RG (2016) Microswimmers–from single particle motion to collective behavior. Eur Phys J Spec Top 225:2061–2064. https://link.springer.com/article/10.1140/epjst/e2016-60095-3
22. Daddi-Moussa-Ider A, Löwen H, Liebchen B (2021) Hydrodynamics can determine the optimal route for microswimmer navigation. Commun Phys 4(1):15. https://doi.org/10.1038/s42005-021-00522-6
23. Negrón-Oyarzo I, Espinosa N, Aguilar-Rivera M, Fuenzalida M, Aboitiz F, Fuentealba P (2018) Coordinated prefrontal-hippocampal activity and navigation strategy-related prefrontal firing during spatial memory formation. Proc Natl Acad Sci 115(27):7123–7128. https://doi.org/10.1073/pnas.1720117115
24. Meyer H, Rieger H (2021) Optimal non-markovian search strategies with n-step memory. Phys Rev Lett 127:070601. https://doi.org/10.1103/PhysRevLett.127.070601
25. Zhu J, Hou Y (2019) Optimal control of navigation systems with time delays using neural networks. In: Zhai XB, Chen B, Zhu K (eds) Machine learning and intelligent communications. Springer International Publishing, Cham, pp 735–747. 978-3-030-32388-2
26. Fernandez-Rodriguez MA, Grillo F, Alvarez L, Rathlef M, Buttinoni I, Volpe G, Isa L (2020) Feedback-controlled active brownian colloids with space-dependent rotational dynamics. Nat Commun 11(1):4223. https://doi.org/10.1038/s41467-020-17864-4

27. Haeufle DFB, Bäuerle T, Steiner J, Bremicker L, Schmitt S, Bechinger C (2016) External control strategies for self-propelled particles: optimizing navigational efficiency in the presence of limited resources. Phys Rev E 94:012617. https://doi.org/10.1103/PhysRevE.94.012617
28. Parrondo JMR, Horowitz JM, Sagawa T (2015) Thermodynamics of information. Nat Phys. https://www.nature.com/articles/nphys3230
29. Sartori P, Granger L, Lee CF, Horowitz JM (2014) Thermodynamic costs of information processing in sensory adaptation. PLOS Comput Biol 10(12):1–9. https://doi.org/10.1371/journal.pcbi.1003974
30. Faist P, Dupuis F, Oppenheim J, Renner R (2015) The minimal work cost of information processing. Nat Commun 6(1). https://doi.org/10.1038/ncomms8669

Appendix A
Numerical Methods

A.1 Brownian Dynamics Simulations

The Brownian dynamics simulations of both Eq. (3.5) and Eq. (4.1) have been performed using an Euler-Mayurama scheme with a time step $dt = 10^{-3}$. We have verified that the selected time step is sufficiently small, such that the results presented therein do not depend on its value.

A.1.1 Implementation of the Semi-autonomous Navigation Protocols

Below we provide details regarding the implementation of the navigation strategies presented in Sects. 4.1, 4.4.

In all our simulations a given run ends when the active particle is within a distance $\delta r = 0.025\ell$ from the target. We have also checked that the choice of the disk radius δr does not significantly influence the results as long as the thermal fluctuations length scale is kept relatively small, i.e. $\sqrt{2Ddt} \ll \delta r \ll \ell$.

The MFPT equation (4.2) was numerically solved using the Finite Elements Method implemented in the NDSolve routine of Wolfram Mathematica 13.1.0 [1]. The Optimal Policy was then implemented from the corresponding solution (4.3) by discretizing the simulation domain on a square grid of step $l = 0.01$, and assigning to each box the optimal control orientation $\hat{\boldsymbol{u}}_{\text{opt}}(\boldsymbol{r}_b)$, with $\boldsymbol{r}_b$ being the position of the centre of the box. In the stochastic simulations, the swimmer following OP was then aligning its direction of motion with the orientation associated with its current position on the grid.

All then newly introduced policies, namely AP, AAP, CP, CAP and CAAP, rely on the evaluation of the point $\boldsymbol{r}_c$ on the Zermelo path closest to the particle position $\boldsymbol{r}$. For numerical efficiency, the Zermelo path was thus discretized and the distance

L. Piro, *Optimal Navigation in Active Matter*, Springer Theses,
https://doi.org/10.1007/978-3-031-52577-3

between the particle and the curve was calculated from the positions of the mid-point of each segments. In all simulations the initial particle orientation was taken to be equal to the one prescribed by the Zermelo solution.

A.1.2 Langevin Simulations on the Sphere

To describe the motion of an overdamped particle on the sphere, the Langevin equation (4.1) has to be adjusted to take into account the multiplicative noise induced by the space curvature. Namely, it is given by

$$\dot{\boldsymbol{r}} = v_0\hat{\boldsymbol{u}} + \boldsymbol{f}(\boldsymbol{r}) + \sqrt{2D}\hat{\boldsymbol{r}} \times \boldsymbol{\xi} \,, \tag{A.1}$$

where $\hat{\boldsymbol{r}} \equiv \boldsymbol{r}/|\boldsymbol{r}|$, while the noise $\boldsymbol{\xi}$ shares the same statistics as in Eq. (4.1) and is interpreted in the Stratonovich sense. Furthermore, in contrast with the Taylor-Green flow case studied in Chap. 4, for this setup the characteristic length scale of the flow is comparable with the sphere radius: $\ell \sim R$, such that the Péclet number is here defined as $\mathrm{Pe} = Rv_0/D$.

Lastly, the extension of SP, AP and AAP to the case where motion takes place on a sphere straightforwardly follows from the presentation in the text, as it only requires us to generalize the definitions of the distance and relative direction between two points of interest. On a sphere of radius R, the shortest distance between the points $\boldsymbol{r}$ and $\boldsymbol{r}_c$—also known as great-circle distance– is defined as

$$|\Delta r| \equiv R \arccos\left(\frac{\boldsymbol{r} \cdot \boldsymbol{r}_c}{R^2}\right) .$$

The direction $\hat{\boldsymbol{n}}$ from $\boldsymbol{r}$ to $\boldsymbol{r}_c$ at the point $\boldsymbol{r}$ is likewise defined from the shortest arc linking the two points, namely

$$\hat{\boldsymbol{n}} \equiv \frac{(\boldsymbol{r} \times \boldsymbol{r}_c) \times \boldsymbol{r}}{|(\boldsymbol{r} \times \boldsymbol{r}_c) \times \boldsymbol{r}|} .$$

The desired heading direction $\hat{\boldsymbol{u}}$ for AP and AAP was then obtained from rotations of $\hat{\boldsymbol{n}}$ around the axis set by $\boldsymbol{r}$ using Rodrigues rotation formula [2]:

$$\hat{\boldsymbol{u}} = \mathcal{G}(\Delta r)\hat{\boldsymbol{n}} \pm (\boldsymbol{r} \times \hat{\boldsymbol{n}})\sqrt{1 - \mathcal{G}^2(\Delta r)} + \boldsymbol{r}(\boldsymbol{r} \cdot \hat{\boldsymbol{n}})(1 - \mathcal{G}(\Delta r)) \,,$$

where the protocol function $\mathcal{G}(\Delta r)$ is defined by Eq. (4.4) and the $\pm$ sign ensures that $\hat{\boldsymbol{u}} \cdot \hat{\boldsymbol{t}} \geq 0$.

A.2 Details on the Numerical Simulations of a Gaussian Random Flow Field

The Gaussian random flow presented in Sect. 4.4.2 has been obtained following the power spectrum realization method [3]. First of all, we build a 500×500 grid containing uncorrelated Gaussian random numbers and then Fourier transform the result. In Mathematica [1] this can be done via the following commands

```
WhiteNoise = RandomVariate[NormalDistribution[],{Nsize, Nsize}]
WhiteNoiseFourier = Fourier[WhiteNoise] ,
```

where Nsize $= 500$ is the dimension of the grid. We shall then multiply the outcome with the square root of the power spectrum of the stream function $\psi(\boldsymbol{r})$—the Fourier transform of its correlation function (4.13)—and finally transform back the result using the InverseFourier routine in Mathematica. In order to determine the corresponding velocity field, one simply then has to compute the resulting stream function derivatives via finite differences and use Eq. (4.12).

In our numerical simulations we have set the characteristic length and flow intensity scales to $\ell = 0.6$ and $v_\mathrm{f} = 0.6$, respectively. The flow field was computed in a square box of size $L \times L$ with $L = 25/3\ell$ and comes out naturally with periodic boundary conditions.

A.3 Numerical Implementation of a Multi-dimensional Shooting Method

In order to solve the ODE system (5.23), we have to determine the correct value of the unknown initial conditions for the heading angle θ_0 and the momentum components $p_{x,0}$ and $p_{y,0}$ such that the boundary conditions

$$\begin{cases} \boldsymbol{r}(0) = \boldsymbol{r}_0\,, \quad \boldsymbol{r}(t_\mathrm{f}) = \boldsymbol{r}_\mathrm{T} \\ \omega_a(0) = \omega_a(t_\mathrm{f}) = 0 \end{cases}$$

are met. To this end, we thus had to perform a multi-dimensional shooting method. This has been achieved by using the GSL multi-dimensional root-finding routines [4] included in the header file *gsl_multiroots.h*.

Starting from a random initial condition for θ and $\boldsymbol{p}$, we therefore first integrate the ODE system (5.23) up to a certain time t_f using a fourth order Runge-Kutta method with time step $\mathrm{d}t = 10^{-5}$. We then perform an iteration of the GSL multi-dimensional root-finder which, while looking for the solution of the system

$$\begin{cases} \boldsymbol{r}(t_{\mathrm{f}}) - \boldsymbol{r}_{\mathrm{T}} = 0 \\ \omega_a(t_{\mathrm{f}}) = 0\,, \end{cases} \tag{A.2}$$

updates the unknown initial conditions. This process is then repeated until convergence, i.e. when the residual of each equation in (A.2) is below a specified threshold.

Actually, to speed up the convergence of our simulations, for each value of α we have initially fixed the final time $t_{\mathrm{f}} = t_{\mathrm{p}}$, namely equal to the arrival time without active steering. We could then use our previous knowledge about the correct initial orientation θ_0 to be taken (see results from Sect. 5.2.2). Furthermore, from the equation for ω_a in (5.23), we could also impose the condition $\boldsymbol{p}(0) \cdot \hat{\boldsymbol{u}}^{\perp}(0) = 0$, since in this case we know that ω_a should stay equal to zero throughout the trajectory. Thanks to this refined choice of the initial conditions, the code converged faster and more robustly, such that we could then start from this solution to systematically scan times $t_{\mathrm{f}} > t_{\mathrm{p}}$ or $t_{\mathrm{f}} < t_{\mathrm{p}}$ for each α.

References

1. Wolfram Research (2022) Inc. Mathematica, Version 13.1.0. https://www.wolfram.com/mathematica
2. Rodrigues O (1840) Des lois géométriques qui régissent les déplacements d'un système solide dans l'espace, et de la variation des coordonnées provenant de ces déplacements considérés indépendamment des causes qui peuvent les produire. Journal de Mathématiques Pures et Appliquées 380–440. http://eudml.org/doc/234443
3. Goon G (2021) Gaussian fields. https://garrettgoon.com/gaussian-fields/
4. Gough B (2009) GNU scientific library reference manual. Network Theory Ltd